Statistics for Life Science 2

ULF OLSSON

⚠ **Copying prohibited**

All rights reserved. No part of this publication may be reproduced or transmitted in any form or by any means, electronic or mechanical, including photocopying, recording, or any information storage and retrieval system, without permission in writing from the publisher.

The papers and inks used in this product are eco-friendly.

Art. No 34292
ISBN 978-91-44-07149-7
Edition 1:1

© The author and Studentlitteratur 2011
www.studentlitteratur.se
Studentlitteratur AB, Lund

Cover design by Francisco Ortega
Cover images by kentoh, Raj Creationzs, Ashusha and Sergey150770/Shutterstock

Printed by Holmbergs i Malmö AB, Sweden 2011

Contents

Preface 9

1 Simple linear regression   **11**
    1.1   Introductory example 11
    1.2   Model and assumptions 12
    1.3   Parameter estimates 13
    1.4   ANOVA table and F test 15
    1.5   Tests and confidence intervals for β_1 17
    1.6   Forecasting 17
        1.6.1   A confidence interval for $E(y|x_*)$ 17
        1.6.2   Prediction intervals for one observation with $x = x_*$ 18
    1.7   The coefficient of determination, R^2 19
    1.8   The correlation coefficient ρ when x is random 19
    1.9   Testing linearity 21
    1.10   Regression diagnostics 25
    1.11   Analysis by computer 28
        1.11.1   Analysis using Minitab 28
        1.11.2   Analysis using SAS 29
        1.11.3   Testing linearity 30
    1.12   Appendix 32
        1.12.1   Derivation of least-sqares estimates in linear regression 32
        1.12.2   A proof that $\sum (y_i - \hat{y}_i)(\hat{y}_i - \overline{y}) = 0$ 33
    1.13   Exercises 34

2 Multiple linear regression   **39**
    2.1   Introduction 39
    2.2   The multiple regression model 40
    2.3   Estimation 41
    2.4   Sum of squares decomposition 41

Contents

- 2.5 R^2: an overall measure of fit 42
- 2.6 Inference on single parameters 43
- 2.7 Analysis by computer 43
- 2.8 Tests on subsets of the parameters 44
- 2.9 Prediction 45
- 2.10 Model building 46
  - 2.10.1 Partial and sequential tests 46
  - 2.10.2 Stepwise regression 48
- 2.11 Exercises 50

3 Nonlinear models 53
- 3.1 Models that are linear in the parameters 53
- 3.2 Models that can be linearized by transformation 56
- 3.3 Box-Cox transformations 58
- 3.4 Truly nonlinear models 59
- 3.5 "Transform both sides" 63
- 3.6 Exercises 66

4 General linear models 69
- 4.1 The role of models 69
- 4.2 General linear models 70
- 4.3 Estimation 71
- 4.4 Assessing the fit of the model 71
  - 4.4.1 Predicted values and residuals 71
  - 4.4.2 Sums of squares decomposition 72
- 4.5 Inference on single parameters 73
- 4.6 Tests on subsets of the parameters 74
- 4.7 Different types of tests 75
- 4.8 Some applications 76
  - 4.8.1 Simple linear regression 76
  - 4.8.2 Multiple regression 77
  - 4.8.3 t tests and dummy variables 79
  - 4.8.4 One-way ANOVA 81
  - 4.8.5 ANOVA: Factorial experiments 85
  - 4.8.6 Analysis of covariance 88
- 4.9 Estimability 90
- 4.10 Assumptions in general linear models 91
- 4.11 Model building 91
  - 4.11.1 Computer software for GLMs 91
  - 4.11.2 Model building strategy 92
- 4.12 A covariance analysis example 93
  - 4.12.1 Two-way ANOVA without interaction 93
  - 4.12.2 A model with interaction 95

	4.13 General linear models using matrix algebra	97
	4.14 GLM: Summary	101
	4.15 Exercises	102

5 Nonparametric methods — 109

- 5.1 One sample: the sign test ... 109
- 5.2 Two independent samples: the Wilcoxon-Mann-Whitney test ... 111
- 5.3 More than two samples: the Kruskal-Wallis test ... 113
- 5.4 Randomized block designs ... 115
  - 5.4.1 Two samples: the sign test ... 115
  - 5.4.2 More than two samples in randomized blocks: Friedman's test ... 116
- 5.5 Rank transformations: ANOVA on ranks ... 117
- 5.6 Randomization tests ... 119
  - 5.6.1 Fisher's exact test ... 119
  - 5.6.2 Other applications of randomization tests ... 120
- 5.7 Some comments on nonparametric tests ... 122
- 5.8 Exercises ... 124

6 Analysis of frequency data — 127

- 6.1 Test of model fit ... 127
  - 6.1.1 Testing model fit: the distribution is completely specified ... 127
  - 6.1.2 Merging categories ... 129
  - 6.1.3 Test of model fit: the parameters are unknown ... 130
- 6.2 Testing homogeneity ... 132
- 6.3 Testing independence ... 134
- 6.4 Some comments on χ^2 tests ... 138
- 6.5 Exercises ... 141

7 Generalized linear models — 145

- 7.1 Introduction ... 145
- 7.2 Generalized linear model theory ... 146
  - 7.2.1 The exponential family of distributions ... 147
- 7.3 Analysis of binary data ... 148
  - 7.3.1 Logistic regression ... 149
  - 7.3.2 Odds ratios ... 152
  - 7.3.3 Multiple logistic regression ... 153
- 7.4 Analysis of counts: log-linear models ... 157
  - 7.4.1 Two-way contingency tables ... 157
  - 7.4.2 A log-linear model for independence ... 157
  - 7.4.3 When independence does not hold ... 158

Contents

- 7.4.4 Distributions for count data 159
- 7.4.5 Relation to contingency tables 160
- 7.4.6 Analysis of the example data 161
- 7.4.7 Higher-order tables 162
- 7.4.8 Types of independence 162
- 7.4.9 Genmod analysis of the drug use data 163
- 7.4.10 Interpretation through odds ratios 164
- 7.5 Poisson regression 165
- 7.6 Rate data 167
- 7.7 Ordinal data 169
- 7.8 Gamma distribution 172
- 7.9 Over-dispersion 174
- 7.10 Exercises 178

8 Introduction to repeated-measures data   183
- 8.1 An example 183
- 8.2 Graphical displays for repeated-measures data 184
  - 8.2.1 Individual profile plot 185
  - 8.2.2 Mean profile plot 185
  - 8.2.3 Boxplots 186
- 8.3 Historical approaches 186
  - 8.3.1 The "summary-measures" approach 187
  - 8.3.2 Repeated-measures ANOVA 189
  - 8.3.3 Growth curve analysis using MANOVA 192
- 8.4 Analysis as a mixed linear model 194
  - 8.4.1 Introduction to mixed models 194
  - 8.4.2 A model with compound symmetry 195
  - 8.4.3 Modeling the covariance structure 197
  - 8.4.4 Using the "repeated" statement 198
  - 8.4.5 Further improvements of the model 199
  - 8.4.6 A strategy for mixed model analysis 203
- 8.5 Analysis of Cross-over data 204
  - 8.5.1 Introduction 204
  - 8.5.2 Designs 205
  - 8.5.3 Mixed models 206
- 8.6 Exercises 211

9 Introduction to multivariate methods   213
- 9.1 Describing multivariate data 213
  - 9.1.1 The data matrix 213
  - 9.1.2 Uses of multivariate statistics 214
  - 9.1.3 The mean vector and the covariance matrix ... 214
- 9.2 Principal component analysis 216

	9.2.1	Introduction	216
	9.2.2	Definition of principal components	216
	9.2.3	Explained variance	218
	9.2.4	Analysis of the example data	218
	9.2.5	Number of components to retain	220
9.3	Factor analysis		222
9.4	Cluster analysis		225
	9.4.1	Measures of similarity or distance	225
	9.4.2	Measures of "distance" between clusters	226
	9.4.3	Clustering variables	228
	9.4.4	Other clustering methods	229
	9.4.5	Some advice on choice of method	231
9.5	Exercises		232

10 Topics in clinical trials   233

10.1	Introduction to clinical trials		233
	10.1.1	Terminology and brief historical overview	233
	10.1.2	Methods of investigation in clinical research	234
	10.1.3	Methods for clinical trials	238
	10.1.4	Phases of clinical trials	239
	10.1.5	Stages of a clinical trial	239
10.2	Design of clinical trials		240
	10.2.1	The purpose of a clinical trial	240
	10.2.2	Variables and endpoints	240
	10.2.3	The study protocol	241
	10.2.4	Parallel groups designs	241
	10.2.5	Blocking	243
	10.2.6	Stratification	245
	10.2.7	Minimization and balancing	246
	10.2.8	Cross-over trials	246
10.3	Reliability and validity		247
	10.3.1	A measurement model	248
	10.3.2	Effects of measurement errors	248
	10.3.3	Estimating the reliability	250
10.4	Early stopping		252
	10.4.1	Methods to make conclusions during the trial	252
	10.4.2	Alpha spending functions	254
10.5	Equivalence studies		255
	10.5.1	"Clinically important differences"	257
	10.5.2	Superiority and inferiority	259
	10.5.3	Equivalence studies with proportions	259
	10.5.4	Sample size determination	260
10.6	Dose-finding studies		261

Contents

        10.6.1 Dose-response studies 261
        10.6.2 Dose-finding studies 262
        10.6.3 A Bayesian approach: Continual reassessment . . . 263
        10.6.4 Storer's two-stage design 265
  10.7 Exercises 266

Tables   **271**

Appendix A: Introduction to matrix algebra   **287**
  A.1 Some basic definitions 287
  A.2 The dimension of a matrix 288
  A.3 The transpose of a matrix 288
  A.4 Some special types of matrices 288
  A.5 Calculations on matrices 289
  A.6 Matrix multiplication 290
        A.6.1 Multiplication by a scalar 290
        A.6.2 Multiplication by a matrix 290
        A.6.3 Calculation rules of multiplication 291
        A.6.4 Idempotent matrices 291
  A.7 The inverse of a matrix 291
  A.8 Generalized inverses 292
  A.9 The rank of a matrix 292
  A.10 Determinants 293
  A.11 Eigenvalues and eigenvectors 293
  A.12 Some statistical formulas on matrix form 294
  A.13 Further reading 294

Bibliography   **295**

Answers to the exercises   **305**

Index   **337**

Preface

This is the second of two books on "Statistics for life". It has grown out of material prepared for the graduate courses Statistics for Biologists I and Statistics for Biologists II that have been given at the Swedish University of Agricultural Sciences for a number of years. The material has also been used in various statistics courses for students in medicine, nursing and pharmacology at Uppsala University.

The book is intended as course material for courses in statistics for students majoring in a biological subject. We have used it for a 7.5 ECTS credit course for students in the second half of their undergraduate studies, or at the early stage of their Master/Ph.D. studies, who have already taken courses corresponding to "Statistics for Life Science I".

The emphasis is on methods for drawing conclusions from biological data. Some special features of the book are

- It includes some advanced topics that are not normally included in introductory texts, such as general linear models, generalized linear models, and mixed models for analysis of repeated-mesures data.

- We assume that the reader will solve many of the exercises using a computer.

- Examples of SAS code for solving some of the exercises are included.

The book starts with chapters on simple and multiple linear regression. The concept of general linear models, including covariance analysis, is introduced through the use of dummy variables. General nonlinear regression models are discussed. Methods for non-normal data are covered in several chapters: one chapter on nonparametric methods; one chapter on analysis of contingency tables, and one chapter on generalized linear models. The treatment of the analysis of repeated-measures data is

Contents

based on the concept of mixed linear models. A chapter on multivariate methods introduces methods such as principal component analysis, factor analysis and cluster analysis. Finally, some topics relevant for clinical trials are discussed such as early stopping, equivalence studies and dose-finding studies.

Most of the applications use real data from actual research. The analyses are illustrated with printouts from the SAS and Minitab packages. Each chapter includes a number of exercises with solutions. Supplementary material is available at the book's home page.

Many people have contributed in different ways to the work with these books. My late colleague Ulla Engstrand was a good friend, as well as coauthor of two previous books. Other friends and colleagues who have supported the work during the years include Razaw Al-Sarraj, Mikael Andersson, Gunnar Ekbohm, Jan-Eric Englund, Anna Gunsjö, Elizabeth Hillerius, Bo Lindqvist, Behnaz Mazogi, Lennart Norell, Birgitta Vegerfors-Persson and Claudia von Brömssen. Many students have helped to detect errors or question marks in previous editions. Gunnar Hjert has competently revised my English. A warm thanks to all of you.

Uppsala in June 2011
Ulf Olsson

Chapter 1

Simple linear regression

Regression analysis is used to study the relation between a quantitative dependent variable (response variable) y and one or more quantitative variables x_1, x_2,...,x_p. The variables x_1, x_2,...,x_p are sometimes called "independent" variables. This does not imply that they are independent of each other, but that we assume that y depends on the x variables. Some simple examples: y might be yield, x_1 is the amount of rain and x_2 is temperature. Or y might be milk yield, x_1 is the age of the cow and x_2 is the weight of the cow. The purpose of the analysis is to find models that are simple, but still fit well with the data. The models can be linear or non-linear. A common aim with regression models is to predict the value of y for given values of the independent variables x_1, x_2,...,x_p. In this chapter, we will study in some detail the simple case where there is only one x variable and where the relation between x and y is linear, apart from random variation.

1.1 Introductory example

Example 1.1 *Zagal et al (1993) made an experiment where the emission of CO_2 from the root system of barley was studied. The emitted amount of CO_2 was measured on a number of plants at different times after germination. New plants were used for each time point. The results are summarized in Table 1.1.*

The purpose of the experiment is to study how $y=CO_2$ emission develops over time. In this experiment, the values of x were chosen beforehand, and they are therefore "fixed". In other cases x might be something that has been measured, outside the control of the experimenter. In such cases, x is a random factor.

1 Simple linear regression

Table 1.1: Relation between time and emission of carbon dioxide from barley roots.

	\multicolumn{4}{c}{x=time}			
	24	30	35	38
y	15.255	28.200	32.862	41.677
	11.069	26.765	34.730	43.448
	10.481	28.414	35.830	45.351

A possible analysis of these data is to make a one-way ANOVA with $a = 4$ levels of the factor "time". This analysis might show that there are differences among the time points, but it does not use the fact that "time" is quantitative. A graph of the relation between x and y is given in Figure 1.1.

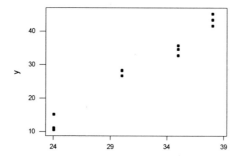

Figure 1.1: The relation between CO_2 emission and time after planting.

The graph in Figure 1.1 suggests that the relation between x and y is rather linear for the range of values of x that are used in the experiment. It seems to be a good idea to fit a straight line to the data.

1.2 Model and assumptions

A model that can be used to summarize our data is

$$y_i = \beta_0 + \beta_1 x_i + e_i.$$

We assume that the residuals e_i are independent, normally distributed with mean zero, and that they have the same variance σ_e^2 for all x. This

means that
$$E(y|x) = \beta_0 + \beta_1 x.$$
This is the equation for a straight line where the intercept is β_0, and where the line has slope β_1. The slope β_1 indicates the expected change in y for a change in x of one unit. If the time increases with one day, the CO_2 emission would increase with β_1 units, on the average. Formally, β_0 can be interpreted as the value of y when $x = 0$. Since the time point $x = 0$ is not part of our data and is not even close to the data, we should not expect β_0 to have any realistic interpretation for our data.

1.3 Parameter estimates

The parameters β_0 and β_1 are estimated using the method of least squares. This means that we place the line where the sum of the squared distances between the points and the line is as small as possible. It can be shown (see the chapter appendix for details) that this gives the following estimators of β_0 and β_1:

$$\widehat{\beta}_1 = b_1 = \frac{\sum_{i=1}^{n}(x_i - \overline{x})(y_i - \overline{y})}{\sum_{i=1}^{n}(x_i - \overline{x})^2}$$

$$\widehat{\beta}_0 = b_0 = \overline{y} - b_1 \overline{x}$$

We will often denote estimators of parameters with a "hat", $\widehat{}$. Thus, $\widehat{\beta}_1$ denotes an estimator of β_1.

For hand calculation it is more convenient to use the following formulas. We define

$$SP_{xy} = \sum_{i=1}^{n}(x_i - \overline{x})(y_i - \overline{y}) = \sum x_i y_i - \frac{\sum x_i \sum y_i}{n}$$

and

$$SS_x = \sum_{i=1}^{n}(x_i - \overline{x})^2 = \sum x_i^2 - \frac{\left(\sum x_i\right)^2}{n}.$$

1 Simple linear regression

Then,

$$b_1 = \frac{SP_{xy}}{SS_x}$$
$$b_0 = \bar{y} - b_1 \bar{x}$$

For later calculations we also need

$$SS_y = \sum_{i=1}^{n}(y_i - \bar{y})^2 = \sum y_i^2 - \frac{\left(\sum y_i\right)^2}{n}.$$

The equation of the curve is estimated as

$$\hat{y} = \hat{\beta}_0 + \hat{\beta}_1 x = b_0 + b_1 x.$$

The calculations for the data in Example 1.1 are summarized in Table 1.2.

Table 1.2: Calculations for a regression analysis of the CO_2 data.

x	y	x^2	y^2	xy
24	15.255	576	232.715025	366.120
24	11.069	576	122.522761	265.656
24	10.481	576	109.851361	251.544
30	28.200	900	795.240000	846.000
30	26.765	900	716.365225	802.950
30	28.414	900	807.355396	852.420
35	32.862	1225	1079.911044	1150.170
35	34.730	1225	1206.172900	1215.550
35	35.830	1225	1283.788900	1254.050
38	41.677	1444	1736.972329	1583.726
38	43.448	1444	1887.728704	1651.024
38	45.351	1444	2056.713201	1723.338
381	354.082	12435	12035.336846	11962.548

We find

$$SP_{xy} = 11962.548 - \frac{381 \cdot 354.082}{12} = 720.445.$$

$$SS_x = 12435 - \frac{381^2}{12} = 338.250.$$

$$SS_y = 12035.336846 - \frac{354.082^2}{12} = 1587.498.$$

Furthermore, $\bar{x} = \frac{381}{12} = 31.75$ and $\bar{y} = \frac{354.082}{12} = 29.507$. This gives

$$b_1 = \frac{720.44}{338.25} = 2.13$$

and

$$b_0 = 29.507 - 2.13 \cdot 31.75 = -38.12.$$

The equation of the estimated line is

$$\hat{y} = -38.12 + 2.13x.$$

Minitab can draw the line together with the data points; see Figure 1.2.

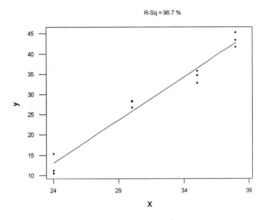

Figure 1.2: Relation between CO_2 emission and time: data points and fitted regression line.

1.4 ANOVA table and F test

In the same way as in ANOVA, the variation in the data can be subdivided into parts:

$$\begin{aligned} SS_T &= \sum_i (y_i - \bar{y})^2 = \sum_i (y_i - \hat{y}_i + \hat{y}_i - \bar{y})^2 \\ &= \sum_i (\hat{y}_i - \bar{y})^2 + \sum_i (y_i - \hat{y}_i)^2 + 2\sum_i (y_i - \hat{y}_i)(\hat{y}_i - \bar{y}). \end{aligned}$$

1 Simple linear regression

The last term can be shown to be equal to zero; see the chapter appendix. The subdivision can symbolically be written

$$SS_T = SS_R + SS_e.$$

The total sum of squares, SS_T, is defined in the same way as in ANOVA:

$$SS_T = SS_y = \sum_i (y_i - \bar{y})^2.$$

$SS_R = \sum_i (\hat{y}_i - \bar{y})^2$ is called the regression sum of squares. It has one degree of freedom. An easy way to calculate SS_R is

$$SS_R = \frac{(SP_{xy})^2}{SS_x} = b_1 \cdot SP_{xy}$$

The residual sum of squares is

$$SS_e = \sum_i (y_i - \hat{y}_i)^2 = \sum_i (y_i - b_0 - b_1 x_i)^2 = SS_T - SS_R$$

with $n - 2$ degrees of freedom.

The results can be summarized in an ANOVA table. This can be used to investigate if there are any linear relationships between x and y.

Source	d.f.	SS	MS	E(MS)
Regression	1	SS_R	MS_R	$\sigma_e^2 + \beta_1^2 \sum_i (x_i - \bar{x})^2$
Residual	$n-2$	SS_e	$MS_e = s_e^2$	σ_e^2
Total	$n-1$	SS_T		

The hypothesis H_0: $\beta_1 = 0$, i.e. that there is no linear relation between x and y, can be tested against the alternative hypothesis H_1: $\beta_1 \neq 0$ using

$$F_{1;n-2} = \frac{MS_R}{s_e^2}.$$

For the data in Example 1.1 it holds that $\hat{y} = -38.12 + 2.13x$. The sums of squares are obtained as $SS_R = 1534.5$ and $SS_T = 1587.5$ from which $SS_e = 1587.5 - 1534.5 = 53.0$. The ANOVA table is:

Source	d.f.	SS	MS
Regression	1	1534.5	1534.5
Residual	10	53.0	5.3
Total	11	1587.5	

We can test H_0: $\beta_1 = 0$ using $F = \frac{1534.5}{5.3} = 289.5$ with $(1; 10)$ d.f. This is highly significant ($p < 0.001$).

1.5 Tests and confidence intervals for β_1

The variance of b_1 is

$$Var(b_1) = \frac{\sigma_e^2}{\sum_i (x_i - \bar{x})^2}.$$

It can be estimated as

$$\widehat{Var}(b_1) = \frac{s_e^2}{\sum_i (x_i - \bar{x})^2}$$

where $s_e^2 = MS_e$ with $n - 2$ d.f. We denote the standard error of b_1 with $SE_{b_1} = \sqrt{\widehat{Var}(b_1)}$. A confidence interval for the slope β_1 can be computed as

$$b_1 \pm t_{1-\alpha/2;\ n-2} SE_{b_1}.$$

For the data in Example 1.1 we obtain a 95% confidence interval for β_1 as

$$2.13 \pm 2.228 \sqrt{\frac{5.3}{338.25}};$$

$$2.13 \pm 0.28.$$

In a similar way we can test hypotheses about β_1.

H_0: $\beta_1 = 0$ is tested using

$$t_{n-2} = \frac{b_1 - 0}{SE_{b_1}}.$$

This test gives the same p value as the F test we presented above. The t test and the F test are equivalent. It holds that $t_{n-2}^2 = F_{1;\ n-2}$.

1.6 Forecasting

1.6.1 A confidence interval for $E(y|x_*)$

It is sometimes of interest to assess the uncertainty in the position of the line for a given value of x, say x_*. When $x = x_*$ the position of the line

1 Simple linear regression

is estimated as $y_* = b_0 + b_1 x_*$. It holds that

$$SE_{y_*} = \sqrt{s_e^2 \left(\frac{1}{n} + \frac{(x_* - \bar{x})^2}{\sum_i (x_i - \bar{x})^2} \right)}.$$

This can be used to present the uncertainty of the line as a confidence interval for the "true" position of the line at $x = x_*$. The interval can be written $b_0 + b_1 x_* \pm t_{1-\alpha/2;\ n-2} SE_{y_*}$, i.e.

$$b_0 + b_1 x_* \pm t_{1-\alpha/2;\ n-2} \sqrt{s_e^2 \left(\frac{1}{n} + \frac{(x_* - \bar{x})^2}{\sum_i (x_i - \bar{x})^2} \right)}.$$

This is a confidence interval for $E(y|x_*)$. In Example 1.1, suppose that we want to give a confidence interval for $E(y|x_* = 36)$, i.e. for the position of the line for $x = 36$. We obtain the point estimate $y_* = -38.12 + 2.13 \cdot 36 = 38.56$. The confidence interval is

$$38.56 \pm 2.228 \sqrt{5.3 \left(\frac{1}{12} + \frac{(36 - 31.75)^2}{338.25} \right)};\quad 38.56 \pm 1.90.$$

1.6.2 Prediction intervals for one observation with $x = x_*$

Sometimes you want to predict the result of one future observation that has $x = x_*$. This prediction would be made as $y = b_0 + b_1 x_*$. The probability statement we will make is called a prediction interval. It is a measure of the uncertainty of one single observation.

The point estimator is the same as before: $b_0 + b_1 x_*$. To measure the uncertainty we add two components: the uncertainty in the position of the line, and the variation of single observations around the line. This latter variance is equal to σ_e^2. Thus,

$$\widehat{Var}(\text{prediction at } x = x_*) = s_e^2 + SE_{y_*}^2$$
$$= s_e^2 \left(1 + \frac{1}{n} + \frac{(x_* - \bar{x})^2}{\sum_i (x_i - \bar{x})^2} \right).$$

We use the notation $SE_{pred} = \sqrt{\widehat{Var}(\text{prediction at } x = x_*)}$. A prediction interval can be calculated as

$$b_0 + b_1 x_* \pm t_{1-\alpha/2;\, n-2} \cdot SE_{pred}$$

$$b_0 + b_1 x_* \pm t_{1-\alpha/2;\, n-2} \sqrt{s_e^2 \left(1 + \frac{1}{n} + \frac{(x_* - \bar{x})^2}{\sum_i (x_i - \bar{x})^2}\right)}.$$

Suppose that we want a 95% prediction interval for an observation with $x = 36$, for the data in Example 1.1. The interval is

$$38.56 \pm 2.228 \sqrt{5.3 \left(1 + \frac{1}{12} + \frac{(36 - 31.75)^2}{338.25}\right)}; \quad 38.56 \pm 5.47.$$

1.7 The coefficient of determination, R^2

A measure of the goodness of fit of the model to data is given by the coefficient of determination,

$$R^2 = \frac{SS_{Model}}{SS_T} = \frac{SS_R}{SS_T}.$$

R^2 indicates how large a part of the total variation in y can be accounted for by the regression model. It holds that $0 \leq R^2 \leq 1$. A value of R^2 close to 0 means that the linear relation between x and y is weak, while a value close to 1 indicates a strong relationship. For the data in Example 1.1, $R^2 = \frac{1534.5}{1587.5} = 0.967$ which is rather high; there is a strong relationship between x and y.

1.8 The correlation coefficient ρ when x is random

In models where the values of x are random, a useful description of the data is given by the regression line and the correlation coefficient ρ. We define

$$\rho = \frac{\sigma_{xy}}{\sigma_x \sigma_y}.$$

In the "random x" model, $\beta_1 = \frac{\sigma_{xy}}{\sigma_x^2}$. The correlation coefficient ρ indicates the degree of linear relation between y and x. If $\beta_1 = 0$, $\rho = 0$ and vice versa.

1 Simple linear regression

ρ is estimated as

$$r = \frac{\sum_i (x_i - \bar{x})(y_i - \bar{y})}{\sqrt{\sum_i (x_i - \bar{x})^2 \cdot \sum_i (y_i - \bar{y})^2}} = \frac{SP_{xy}}{\sqrt{SS_x \cdot SS_y}}.$$

It holds that $-1 \leq \rho \leq 1$. The same relation holds for the estimator r: $-1 \leq r \leq 1$. A value of r that is numerically large indicates a strong relation between the variables x and y. For models with a single x variable it holds that $r = \sqrt{R^2}$.

The test of H_0: $\rho = 0$ is the same as the test of H_0: $\beta_1 = 0$ that was given on page 17.

A few warnings concerning correlations may be needed:

1. A high correlation between two variables does not necessarily mean that one variable affects the other. A classical example: a high correlation has been demonstrated between the number of stork nests in an area and the fertility rate of the area. This does not mean that storks bring the babies. A probable explanation may be that storks nest in rural areas, which also have higher fertility rates.

2. A correlation between x and y near 0 does not mean that x and y are unrelated. Correlation only measures the degree of *linear* relationship. In fact, the correlation between x and y for the data in Figure 1.3 is zero, although there is a perfect quadratic relation between the variables.

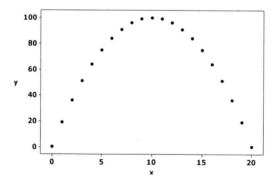

Figure 1.3: A set of data with perfect relation between x and y, but where the correlation $r_{xy} = 0$.

Table 1.3: Anscombe's data: we study the relation between x and the three variables y1, y2 and y3, and between x2 and y4.

x	y_1	y_2	y_3	x_2	y_4
10	8.04	9.14	7.46	8	6.58
8	6.95	8.14	6.77	8	5.76
13	7.58	8.74	12.74	8	7.71
9	8.81	8.77	7.11	8	8.84
11	8.33	9.26	7.81	8	8.47
14	9.96	8.10	8.84	8	7.04
6	7.24	6.13	6.08	8	5.25
4	4.26	3.10	5.39	8	5.56
12	10.84	9.13	8.15	8	7.91
7	4.82	7.26	6.42	8	6.89
5	5.68	4.74	5.73	19	12.50

Example 1.2 *Anscombe (1973) presented a few sets of data that illustrate the importance of graphing your data. Anscombe's data are presented in Table 1.3.*

In Anscombe's data the correlations between x and the three variables y_1, y_2 and y_3, and the correlation between x_2 and y_4, are all equal: $r_{xy_1} = r_{xy_2} = r_{xy_3} = r_{x_2y_4} = 0.816$. Thus, all relations are rather strong. However, if you plot the relations, (see Figure 1.4), they look rather different.

The relation between x and y_1 seems to be rather linear, apart from some random variation. The relationship between x and y_2 is non-linear. The relation between x and y_3 is linear, except for one "outlier". For the variable y_4, finally, there is nearly no relationship between x and y: the correlation is caused by one single "outlier". This example illustrates the need to plot the data before any relationships are interpreted.

1.9 Testing linearity

An interesting question in regression analysis is to assess whether the model, for example a linear regression model, is adequate. This issue can sometimes be settled by inspecting a plot of the data. However, in the case where we have several y values for each value of x, it is also possible to formally test the hypothesis that the model is correct.

We would like to distinguish between the two situations illustrated in Figure 1.5 and Figure 1.6.

In Figure 1.5 the mean values of y for each value of x are close to the

1 Simple linear regression

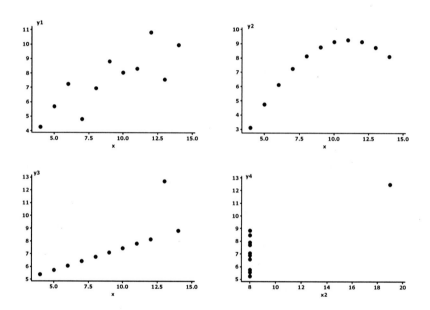

Figure 1.4: Four sets of data where the correlation between x and y is 0.816. Data from Anscombe (1973).

Testing linearity

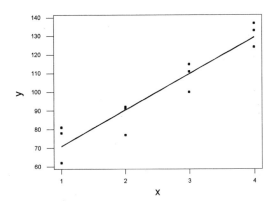

Figure 1.5: Data with a possible linear relation between x and y.

fitted line: the model seems adequate. In Figure 1.6 the mean values seem to systematically deviate from a straight line. A linear model does not seem to fit well.

As a means to test the hypothesis that the "true" regression line is linear, we use the identity

$$(y_{ij} - \overline{y}) = (\widehat{y}_i - \overline{y}) + (\overline{y}_i - \widehat{y}_i) + (y_{ij} - \overline{y}_i).$$

We square and sum over both i and j. It can be shown that all cross-product terms vanish. We get:

$$\sum_{i,j}(y_{ij} - \overline{y})^2 = \sum_{i,j}(\widehat{y}_i - \overline{y})^2 + \sum_{i,j}(\overline{y}_i - \widehat{y}_i)^2 + \sum_{i,j}(y_{ij} - \overline{y}_i)^2.$$

The total sum of squares has been subdivided into the following parts:

$\sum_{i,j}(\widehat{y}_i - \overline{y})^2$ is the familiar regression sum of squares.

$\sum_{i,j}(\overline{y}_i - \widehat{y}_i)^2$ is large if the mean values $\overline{y}_i$ are far from the line (i.e. far from $\widehat{y}_i$). Let us call this the *lack of fit* sum of squares.

$\sum_{i,j}(y_{ij} - \overline{y}_i)^2$ measures the variation around the mean value for each x. It does not depend on the linear model. Let us call this the *pure error* sum of squares.

1 Simple linear regression

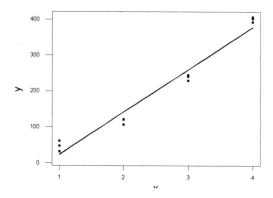

Figure 1.6: Data where the relation between x and y may be nonlinear.

Note that the pure error SS is the same as the error sum of squares if we make a one-way ANOVA on the data, with x as a qualitative factor. Also note that the lack of fit SS plus the pure error SS added together are equal to the residual SS for the regression model.

To test the hypothesis of linearity, we compile an ANOVA table. Here, n is the total number of observations and k is the number of different values of x.

Source	d.f.	SS	MS
Regression	1	$\sum_{i,j} (\widehat{y}_i - \overline{y})^2$	$\sum_{i,j} (\widehat{y}_i - \overline{y})^2 = MS_R$
Lack of fit	$k-2$	$\sum_{i,j} (\overline{y}_i - \widehat{y}_i)^2$	$\dfrac{\sum_{i,j} (\overline{y}_i - \widehat{y}_i)^2}{k-2} = MS_{LOF}$
Pure error	$n-k$	$\sum_{i,j} (y_{ij} - \overline{y}_i)^2$	$\dfrac{\sum_{i,j} (y_{ij} - \overline{y}_i)^2}{n-k} = MS_{PE}$
Total	$n-1$	$\sum_{i,j} (y_{ij} - \overline{y})^2$	

To test the hypothesis that the relation is linear, we calculate the F

ratio

$$F = \frac{\frac{\sum_{i,j}(\overline{y}_i - \widehat{y}_i)^2}{k-2}}{\frac{\sum_{i,j}(y_{ij} - \overline{y}_i)^2}{n-k}} = \frac{MS_{LOF}}{MS_{PE}}$$

which, under H_0, is distributed as F on $(k-2;\ n-k)$ degrees of freedom. An example of this procedure, for our example data, is given on page 30.

1.10 Regression diagnostics

The assumptions underlying a regression analysis are:

The residuals e_i are normally distributed.

The variance of the residuals is constant, independent of x.

The residuals are independent.

The values of x have been measured without error.

In addition, we assume that the model is correct. The first three assumptions are the same as in ANOVA.

Note that the assumptions concern the residuals and not the raw data themselves. To check the assumptions, estimated residuals are calculated as

$$\widehat{e}_i = y_i - \widehat{y}_i$$

where $\widehat{y}_i = b_0 + b_1 x_i$ is the value of y predicted by the model, the so-called fitted value, and y_i is the value actually observed. Most computer packages have an option to calculate and store estimated residuals and fitted values.

Normality

A useful tool for checking the normality assumption is to do a normal probability plot. This is a plot of estimated residuals against their order statistics. If the normality assumption is tenable, the plot should be near linear. In any case, the points should not exhibit a systematic curved pattern. It is possible to make a formal test of normality; however, most normality tests are rather weak, so the plot may be regarded as the main tool for examining normality. Alternatively, histograms or other plots of the residuals may be prepared. An example of a normal probability plot,

1 Simple linear regression

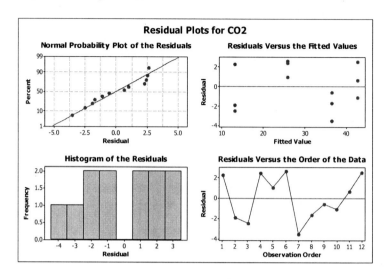

Figure 1.7: Residual plots for the CO_2 data.

based on the data in Example 1.1 is presented in the top left corner of Figure 1.7. A histogram of the residuals is given in the lower left corner. The data do not show any alarming departures from normality.

Homoscedasticity

The model assumes that the variance of the residuals is constant, independent of x. This is called homoscedasticity. A useful way to check this assumption is to prepare a plot of estimated residuals ($\hat{e}$) against fitted values ($\hat{y}$). The graph in the top right corner of Figure 1.7 is an example of this type of plot.

If the variance is not constant, the residuals often display a typical "trumpet shape", i.e. the variance increases with increasing $\hat{y}$. If, on the other hand, the variance is nearly constant, the plot of $\hat{e}$ against $\hat{y}$ does not display any discernible pattern. Figure 1.8 gives an example of residual plot for data that may be homoscedastic while Figure 1.9 is based on heteroscedastic data.

Independence

Diagnosing lack of independence may be rather difficult. However, if there is some natural order in the data, such as a time sequence, it may be useful to plot estimated residuals against order. An example is given

Regression diagnostics

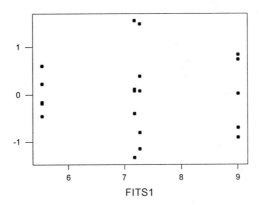

Figure 1.8: Plot of residuals against fitted values for data which may be homoscedastic.

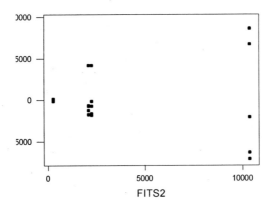

Figure 1.9: Plot of residuals against fitted values for data that do not seem to be homoscedastic.

1 Simple linear regression

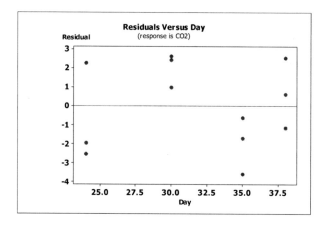

Figure 1.10: Plot of residuals against x for the CO_2 data.

in the lower right corner of Figure 1.7. Departures from independence may be visible as long sequences of residuals with the same sign.

Plots for model checking

In addition to the plots discussed above, it is often useful to plot estimated residuals versus x. This plot can be used to detect possible non-linearities in the data. For our data, this plot is given in Figure 1.10. Non-linearities may be detected as systematic patterns in this kind of plot.

1.11 Analysis by computer

1.11.1 Analysis using Minitab

In Minitab, a regression analysis is obtained via the menu Stat, Regression, Regression. We give y as "response" and x as "model" and get the printout

Regression Analysis

```
The regression equation is
y = - 38.1 + 2.13 x

Predictor      Coef       StDev         T          P
Constant    -38.118       4.030     -9.46      0.000
x            2.1299      0.1252     17.01      0.000

S = 2.302     R-Sq = 96.7%      R-Sq(adj) = 96.3%

Analysis of Variance

Source            DF        SS        MS        F         P
Regression         1    1534.5    1534.5   289.47     0.000
Residual Error    10      53.0       5.3
Total             11    1587.5
```

The equation of the line is given as $\hat{y} = -38.1 + 2.13x$. The relation is strongly significant ($p = 0.000$) and the value of R^2 is large (96.7%). The program gives standard errors for b_0 ("Constant") and b_1 ("x"). It is possible to ask Minitab to compute prediction intervals for values of x_* that are given by the user. Using the "regression plot" Minitab draws the regression line and the observations in the same graph; this graph was presented in Figure 1.2. You can ask for confidence intervals for $E(y|x)$, as well as prediction intervals for single observations. These can be added to the graph, for all values of x. For our data this gives the graph in Figure 1.11.

1.11.2 Analysis using SAS

A regression analysis can be obtained using the SAS procedure GLM:
PROC GLM;
MODEL y = x;
RUN;

Note the difference compared to ANOVA models: we do not give any CLASS variable. SAS will then interpret x as a quantitative (numeric) variable. We get the printout:

1 Simple linear regression

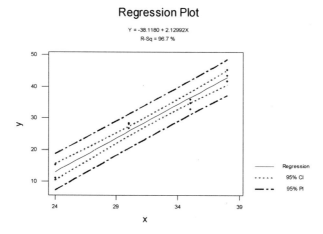

Figure 1.11: Regression line, confidence intervals and prediction intervals for the CO_2 data.

```
General Linear Models Procedure

Dependent Variable: Y

Source              DF        Sum of Squares      Mean Square     F Value     Pr > F
Model                1         1534.48714732     1534.48714732     289.47     0.0001
Error               10           53.01113835        5.30111383
Corrected Total     11         1587.49828567

                R-Square              C.V.            Root MSE              Y Mean

                0.966607           7.802988          2.30241478           29.50683333

Source              DF         Type III SS         Mean Square     F Value     Pr > F

X                    1         1534.48714732     1534.48714732     289.47     0.0001


                                    T for H0:       Pr > |T|     Std Error of
Parameter         Estimate       Parameter=0                         Estimate

INTERCEPT       -38.11803843          -9.46         0.0001         4.02992473
X                 2.12991722          17.01         0.0001         0.12518857
```

There is also a special SAS procedure, Proc REG, that is especially designed for regression analysis.

1.11.3 Testing linearity

To test the hypothesis of linearity, we need to calculate the sums of squares indicated on page 24. Some of the sums of squares can be obtained from the computer outputs above, but in addition we need the output from an ANOVA where x is used as a qualitative factor. This

Analysis by computer

output is as follows:

```
Dependent Variable: y
                                    Sum of
Source                   DF        Squares      Mean Square    F Value    Pr > F
Model                     3    1561.082215       520.360738     157.59    <.0001
Error                     8      26.416071         3.302009
Corrected Total          11    1587.498286

            R-Square      Coeff Var      Root MSE          y Mean
            0.983360       6.158380      1.817143        29.50683

Source                   DF     Type I SS      Mean Square    F Value    Pr > F
x                         3    1561.082215       520.360738     157.59    <.0001
```

We combine the two SAS outputs in the following way:

SS_T: The total sums of squares are the same on the two outputs, $SS_T = 1587.498286$.

SS_{PE}: The pure error sum of squares is the "error" SS from the ANOVA output, i.e. 26.41071 on 8 d.f.

SS_{LOF}: The lack of fit SS is calculated by hand as the difference between the residual SS from the regression, and the pure error SS from the ANOVA: $SS_{LOF} = 53.011113835 - 26.416071 = 26.595043$ on $10 - 8 = 2$ degrees of freedom.

We can compile the ANOVA table:

Source	d.f.	SS	MS
Regression	1	1534.487147	1534.487147
Lack of fit	2	26.595043	13.297521
Pure error	8	26.416071	3.302009
Total	11	1587.498286	

The hypothesis of linearity is tested as $F = \frac{13.297521}{3.302009} = 4.0271$. The 5% limit for F with (2;8) df is 4.459. Although our result is not significant, the fit is not perfect. The p value (calculated by computer) is 0.0617.

1 Simple linear regression

1.12 Appendix

1.12.1 Derivation of least-sqares estimates in linear regression

The estimators of β_0 and β_1 are b_0 and b_1. These are the values that make

$$SS_e = \sum_{i=1}^{n}(y_i - b_0 - b_1 x_i)^2$$

as small as possible, as a function of b_0 and b_1. Note that the values of x_i and y_i are regarded as given constants. The minimum of SS_e is found where the derivatives of SS_e with respect to b_0 and b_1 are zero. These derivatives are

$$\frac{\partial(SS_e)}{\partial b_0} = -2\left(\sum_{i=1}^{n}(y_i - b_0 - b_1 x_i)\right) = 0$$

$$\frac{\partial(SS_e)}{\partial b_1} = -2\left(\sum_{i=1}^{n}(y_i - b_0 - b_1 x_i)x_i\right) = 0$$

The first equation gives

$$\sum_{i=1}^{n}y_i - \sum_{i=1}^{n}b_0 - b_1\sum_{i=1}^{n}x_i = 0.$$

Summation of a constant gives n times the constant, so $\sum_{i=1}^{n}b_0 = nb_0$.

Division by n yields

$$\bar{y} - b_0 - b_1\bar{x} = 0, \text{ i.e.}$$

$$b_0 = \bar{y} - b_1\bar{x}.$$

The second equation gives

$$\sum_{i=1}^{n}x_i y_i - b_0\sum_{i=1}^{n}x_i - b_1\sum_{i=1}^{n}x_i^2 = 0.$$

Insert $b_0 = \bar{y} - b_1\bar{x}$ into this expression :

$$\sum_{i=1}^{n}x_i y_i - n(\bar{y} - b_1\bar{x})\bar{x} - b_1\sum_{i=1}^{n}x_i^2 = 0.$$

Appendix

Simplify with respect to b_1:

$$b_1 \left(\sum_{i=1}^{n} x_i^2 - n\bar{x}^2 \right) = \sum_{i=1}^{n} x_i y_i - n\overline{xy},$$

which gives

$$b_1 = \frac{\sum_{i=1}^{n} x_i y_i - n\overline{xy}}{\sum_{i=1}^{n} x_i^2 - n\bar{x}^2}.$$

This is the calculation formula for b_1 given on page 13. The end result is thus

$$b_1 = \frac{\sum_{i=1}^{n} x_i y_i - n\overline{xy}}{\sum_{i=1}^{n} x_i^2 - n\bar{x}^2}$$

$$b_0 = \bar{y} - b_1 \bar{x}$$

1.12.2 A proof that $\sum (y_i - \hat{y}_i)(\hat{y}_i - \bar{y}) = 0$

$$\sum (y_i - \hat{y}_i)(\hat{y}_i - \bar{y}) =$$

$$\begin{bmatrix} \text{Insert} \\ \hat{y}_i = b_0 + b_1 x_i \end{bmatrix} = \sum (y_i - b_0 - b_1 x_i)(b_0 + b_1 x_i - \bar{y}) =$$

$$\begin{bmatrix} \text{Insert} \\ b_0 = \bar{y} - b_1 \bar{x} \end{bmatrix} = \sum (y_i - \bar{y} + b_1 \bar{x} - b_1 x_i)(\bar{y} - b_1 \bar{x} + b_1 x_i - \bar{y})$$

$$= \sum [(y_i - \bar{y}) - b_1 (x_i - \bar{x})][b_1 (x_i - \bar{x})] =$$

$$= \sum b_1 (x_i - \bar{x})(y_i - \bar{y}) - b_1^2 \sum (x_i - \bar{x})^2 =$$

$$= \begin{bmatrix} \text{Divide both terms with} \\ \sum (x_i - \bar{x})^2 \end{bmatrix} = b_1^2 - b_1^2 = 0 \text{ since}$$

$$b_1 = \frac{\sum (x_i - \bar{x})(y_i - \bar{y})}{\sum (x_i - \bar{x})^2}$$

1 Simple linear regression

1.13 Exercises

Exercise 1.1 *Heggestad and Bennet (1981) studied the effects of air pollution on the yield of Blue Lake snap beans. The beans were grown in field chambers that were fumigated with various levels of sulphur dioxide. After a month of fumigation, the plants were harvested and the yield was recorded. A part of their results was as follows:*

x	SO_2 conc.	0.00	0.00	0.06	0.06	0.12	0.12	0.30	0.30
y	Yield	1.15	1.30	1.19	1.64	1.21	1.00	0.65	0.76

A. *Estimate the coefficients in the linear regression $y = \beta_0 + \beta_1 x + e$.*

B. *Plot the data, and draw the regression line in the graph.*

C. *Compile an ANOVA table and use it to estimate the residual variance s_e^2.*

D. *Test the hypothesis that the level of sulphur dioxide has no linear effect on yield.*

E. *Predict the average (expected) yield when the SO2 concentration is 0.20, and calculate a 95% confidence interval for your prediction.*

Exercise 1.2 *We have the following data on the two fastest times (y, seconds) for running one English mile, for every second year during the period 1954-1972. The plot suggests that a linear trend might be appropriate:*

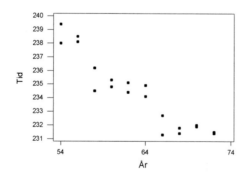

Data: $n = 20$ $\bar{x} = 63$ $\bar{y} = 234.4$ $\sum(x - \bar{x})^2 = 660$
$\sum(x - \bar{x})(y - \bar{y}) = -276$ $\sum(y - \bar{y})^2 = 131.2$.

Fit a straight line to the data. Test H_0: $\beta_1 = 0$. Interpret the parameters β_0 and β_1. Compute R^2.

Can these data be used to predict the best time today?

Exercises

Exercise 1.3 *We study the variables x: a simple method to calculate the amount of ley on an area, and y: actual measurement of the amount (kg dry matter/ha). We have $n = 106$ randomly selected plots. The highest value of x is about 14 and the smallest is about 0.5. A plot of data suggests that a linear trend might be useful. Data: $n = 106$ $\bar{x} = 3.5$ $\bar{y} = 119.1$ $\sum(x-\bar{x})^2 = 1171$ $\sum(x-\bar{x})(y-\bar{y}) = 32393$ $\sum(y-\bar{y})^2 = 943055$.*
Fit a straight line to the data. Test $H_0: \beta_1 = 0$. Interpret the parameters β_0 and β_1. Compute R^2. Calculate and interpret a 95% confidence interval for β_1. Give a point estimate and a prediction interval for y for a new plot for which $x = 10$.

Exercise 1.4 *We study the variables x: height in mm of the shell of a species of snails, and y: width of the shell in mm. We have $n = 28$ randomly selected snails.*
Data: $n = 28$ $\bar{x} = 2.05$ $\bar{y} = 5.4$ $\sum(x-\bar{x})^2 = 3.27$ $\sum(x-\bar{x})(y-\bar{y}) = 6.52$ $\sum(y-\bar{y})^2 = 17.45$.
Fit a linear regression to the data. Test $H_0: \beta_1 = 0$. Interpret the parameters β_0 and β_1. Compute R^2. Estimate the coefficient of correlation ρ.

Exercise 1.5 *When paper pulp is manufactured, the paper will get different properties depending on how long the pulp was mixed. One such property is the so called "Schopper-Riegler freeness test". An experiment was made to measure this property (y) for 13 different batches of paper pulp using different mixing times (x). Results:*

1	2	3	4	5	6	7	8	9	10	11	12	13
17	21	22	27	36	49	56	64	80	86	88	92	94

$$\sum x = 91 \quad \sum x^2 = 819 \quad \sum xy = 6485$$
$$\sum y = 732 \quad \sum y^2 = 51712$$

A. *Estimate the parameters in a simple linear regression of y on x.*
B. *Compile an ANOVA table and use it to test $H_0: \beta_1 = 0$.*
C. *Calculate the residuals and make a graph of their distribution. Is there reason to believe that the residuals are not normal?*
D. *Calculate a 95% confidence interval for the average (expected) value of y, given that $x = 10$.*

Exercise 1.6 *Cicirelli et al (1983) studied protein development in the egg cell of the frog Xenopus laevis. Individual egg cells were injected with radioactively labeled leucine. At different times after injection, radioactivity measurements were made to determine how much of the leucine*

35

1 Simple linear regression

had been incorporated into protein. The results are summarized in the following table:

Time (x)	0	10	20	30	40	50	60
Leucine (y)	0.02	0.25	0.54	0.69	1.07	1.50	1.74

$\bar{x} = 30$ $\quad \sum (x - \bar{x})^2 = 2800 \quad \sum (x - \bar{x})(y - \bar{y}) = 81.90$
$\bar{y} = 0.83$ $\quad \sum (y - \bar{y})^2 = 2.4308$

A. Estimate the parameters in the regression equation $y = \beta_0 + \beta_1 x + e$.

B. Prepare an ANOVA table.

C. Test the hypothesis $H_0: \beta_1 = 0$.

D. Predict the leucine level of a single egg cell at time 45 minutes and give a 95% interval for your prediction.

Exercise 1.7 "Dowsing" is a name for the procedure of finding water using a Y-shaped stick (divining rod; see picture below). A large experiment has been made in Germany (Betz et al, 1990) to investigate whether dowsing actually works. A total of 843 tests were performed, using 43 persons claiming to be "dowsers".

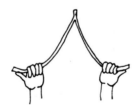

In each test, a water hose was placed at a random location across a line on the first floor of a large barn. The dowser was asked to mark the position of the hose (i.e. to "find water") across a similar line on the second floor of the barn. If the mark were to be "close" to the actual position, this would mean that dowsing works. The following data are the results for one experiment for dowser number 99. The numbers are decimeters from the beginning of the line.

x: pipe location	y: dowser's guess
4	4
5	87
30	95
35	74
36	78
58	65
40	39
70	75
74	32
98	100
$\sum x = 450$	$\sum y = 649$
$\sum x^2 = 28406$	$\sum y^2 = 50565$
$\sum xy = 31447$	

A. Estimate the parameters in a linear regression of y on x.

B. Draw the data and the regression line in the same plot.

C. What is the correlation between y and x?

D. Is the parameter β in the regression significantly different from zero?

E. Assume that dowsing does not work. Out of the 843 experiments, in how many cases would we expect the result to be significant at the 5% level?

Chapter 2

Multiple linear regression

2.1 Introduction

In simple linear regression, we study the relation between a dependent variable y and a single independent variable x. The models are of the form $y = \beta_0 + \beta_1 x + e$. The parameters β_0 and β_1 are estimated with the values that makes the sum of squared residuals as small as possible.

Multiple linear regression generalizes this idea to situations where y is a function of several independent variables. Some examples:

y=Systolic blood pressure of a patient

x_1=Age of the patient

x_2=Weight of the patient

y=Yield of hay per ha

x_1=Total rain amount May-July

x_2=Average temperature May-July

The purpose of the analysis is often to predict the outcome y for specified values of the x variables. In some cases, the purpose is to explain or illustrate a scientific model.

For the computer printouts and analyses in this chapter, the following example will be used:

Example 2.1 *Professor Orley Ashenfelter issues a wine magazine, "Liquid Assets", giving advice about good years. He bases his advice on mul-*

2 Multiple linear regression

Table 2.1: Data for prediction of the quality of wine.

Year	Rain_W	Av_temp	Rain_H	Quality
1975	123	23	23	89
1976	66	21	100	70
1977	58	20	27	77
1978	109	26	33	87
1979	46	22	102	73
1980	40	19	77	70
1981	42	18	85	60
1982	167	25	14	92
1983	99	28	17	87
1984	48	24	47	79
1985	85	24	28	84
1986	177	27	11	93
1987	80	22	45	75
1988	64	25	40	82
1989	75	25	16	88

tiple regression of

$$y = Price\ of\ the\ wine\ at\ wine\ auctions$$

with meteorological data as predictors. The New York Times used the headline "Wine Equation Puts Some Noses Out of Joint" on an article about Prof. Ashenfelter. Base material was taken from "Departures" magazine, September/October 1990, but the data are invented. The variables in the data set in Table 2.1 are:

x_1=Rain_W=Amount of rain during the winter.
x_2=Av_temp=Average temperature.
x_3=Rain_H=Rain in the harvest season.
y=Quality, which is an index based on auction prices.

2.2 The multiple regression model

A multiple regression model with two x variables can be written as

$$y_i = \beta_0 + \beta_1 x_{i1} + \beta_2 x_{i2} + e_i$$

for $i = 1, 2, \ldots, n$. In general, there may be p x variables. The model is then

$$y_i = \beta_0 + \beta_1 x_{i1} + \beta_2 x_{i2} + \ldots + \beta_p x_{ip} + e_i.$$

As usual, we assume that the residuals e_i are independent, normally distributed and that they have the same variance σ_e^2:

$$e_i \sim N\left(0, \sigma_e^2\right).$$

2.3 Estimation

Estimation of parameters in multiple regression models is done using the principle of least squares. This means that we estimate the parameters $\beta_0, \beta_1, \ldots, \beta_p$ with those estimators $b_0, b_1, \ldots, b_p$ for which the sum of the squares of the estimated residuals is as small as possible. The estimators are sometimes denoted $\widehat{\beta}_0$, $\widehat{\beta}_1$, etc.

The predicted value of an observation y_i which has x-values x_{i1}, $x_{i2}, \ldots, x_{ip}$ is

$$\widehat{y}_i = b_0 + b_1 x_{i1} + b_2 x_{i2} + \ldots + b_p x_{ip}.$$

The estimated residual is

$$\begin{aligned}\widehat{e}_i &= y_i - \widehat{y}_i \\ &= y_i - (b_0 + b_1 x_{i1} + b_2 x_{i2} + \ldots + b_p x_{ip}).\end{aligned}$$

The sum of the squared residuals is

$$SS_e = \sum_{i=1}^{n} (y_i - \widehat{y}_i)^2,$$

which can be minimized as a function of $b_0, b_1, \ldots, b_p$. This gives formulas to calculate the estimates, but since we will do the calculations by computer, no hand-calculations formula are presented here.

2.4 Sum of squares decomposition

The performance of a multiple regression analysis can be assessed in the same way as for simple linear regression. We can subdivide the total variation, measured by SS_T, as

$$\begin{aligned}SS_T &= \sum_i (y_i - \overline{y})^2 = \sum_i (y_i - \widehat{y}_i + \widehat{y}_i - \overline{y})^2 \\ &= \sum_i (\widehat{y}_i - \overline{y})^2 + \sum_i (y_i - \widehat{y}_i)^2 + 2\sum_i (y_i - \widehat{y}_i)(\widehat{y}_i - \overline{y}).\end{aligned}$$

2 Multiple linear regression

The expression in the bracket can be shown to be zero (see page 33). As usual, SS_T has $(n-1)$ degrees of freedom. Thus, we make the decomposition

$$SS_T = SS_{\text{Reg}} + SS_e.$$

Here:
$SS_{\text{Reg}} = \sum_i (\widehat{y}_i - \overline{y})^2$ has p degrees of freedom, where p is the number of x variables.

$SS_e = \sum_i (y_i - \widehat{y}_i)^2$ has the remaining degrees of freedom, i.e. $n - p - 1$.

Once the sums of squares have been calculated, we can compile an ANOVA table:

Source	d.f.	SS	MS
Regression	p	SS_{Reg}	MS_{Reg}
Residual	$n-p-1$	SS_e	$MS_e = s_e^2$
Total	$n-1$	SS_T	

It is possible to assess the performance of the regression as a whole by making an over-all F test. This tests the joint hypothesis H_0: $\beta_1 = \beta_2 = \ldots = \beta_p = 0$. It is calculated as

$$F = \frac{MS_{\text{Reg}}}{MS_e}$$

on $(p; n - p - 1)$ degrees of freedom.

2.5 R^2: an overall measure of fit

As a measure of the over-all success of the analysis, we can calculate the coefficient of determination

$$R^2 = \frac{SS_{\text{Reg}}}{SS_T} = 1 - \frac{SS_e}{SS_T}.$$

It holds that $0 \leq R^2 \leq 1$. For a model where the observations are close to the fitted model, the residuals will be small and R^2 will be close to 1, while a model where most variation is random will have a value of R^2 close to 0.

There is a problem with R^2 as a measure of the fit of the model. If n is small, it is always possible to get a seemingly good fit simply by adding more x variables to the model, even if these variables are not related to y. It holds that R^2 will always increase, or be unchanged, when new variables are added to the model. Therefore, an adjusted R^2 has been suggested. This is defined as

$$R^2_{Adj} = 1 - \frac{SS_e/(n-p-1)}{SS_T/(n-1)}.$$

One way to view this is as

$$R^2_{Adj} = 1 - \frac{\text{(Variance estimated from the model)}}{\text{(Variance estimated without any model)}}.$$

The adjusted R^2 is often more useful than the ordinary R^2 for model comparison.

2.6 Inference on single parameters

Expressions for the variances of the estimated coefficients $b_0, b_1,\ldots,b_p$ can be derived. We do not give the formulas here but give reference to computer printouts. These estimated variances can be used to obtain tests or confidence estimates for the parameters.

Denote the estimated variance of b_j with s_j^2. This estimated variance is based on SS_e and thus has $n-p-1$ degrees of freedom. A test of the hypothesis H_0: $\beta_j = 0$ can be obtained as

$$t = \frac{b_j - 0}{s_j}$$

with $n-p-1$ d.f. Similarly, a confidence interval for the parameter β_j at level $(1-\alpha)$ can be calculated as

$$b_j \pm t_{(1-\alpha/2,\ n-p-1)} s_j.$$

2.7 Analysis by computer

A multiple regression output from Minitab based on the data presented in Table 2.1 is as follows:

2 Multiple linear regression

Regression Analysis

```
The regression equation is
Quality = 48.9 + 0.0594 Rain_W + 1.36 Av_temp - 0.118 Rain_H

Predictor        Coef        StDev         T         P
Constant        48.91        10.41       4.70     0.001
Rain_W         0.05937      0.02767      2.15     0.055
Av_temp        1.3603       0.4187       3.25     0.008
Rain_H        -0.11773      0.04010     -2.94     0.014

S = 3.092        R-Sq = 91.6%      R-Sq(adj) = 89.4%

Analysis of Variance

Source          DF         SS         MS         F         P
Regression       3      1152.43     384.14     40.18     0.000
Residual Error  11       105.17       9.56
Total           14      1257.60
```

The output indicates that the three predictor variables do indeed have a relationship to the wine quality, as measured by the price. The variable Rain_W is not quite significant but would be included in a predictive model. The size and direction of this relationship is given by the estimated coefficients of the regression equation. It appears that years with much winter rain, a high average temperature, and only a small amount of rain at harvest time, would produce good wine.

2.8 Tests on subsets of the parameters

Sometimes it may be of interest to test whether a whole group of variables has any significant relation with y. For example, we may want to compare a model that contains the variables x_1, x_2 and x_3 with a model that only contains the variable x_3. This corresponds to a joint test of the hypothesis $H_0: \beta_1 = \beta_2 = 0$ given that $\beta_3 \neq 0$. In general, we may want to compare a model that contains the variables 1, 2, ..., p with a model where the first q parameters are zero. Thus, we test the joint hypothesis $H_0: \beta_1 = \beta_2 = \ldots \beta_q = 0$ given that $\beta_{q+1} \ldots \beta_p$ are different from zero. This is achieved as follows:

1. Make an analysis of the full model with p variables. This will give an error sum of squares, SS_{e1}, with $(n - p - 1)$ degrees of freedom.

2. Estimate the parameters of the smaller model, i.e. the model with fewer parameters. This will give an error sum of squares, SS_{e2}, with $(n - p - q - 1)$ degrees of freedom, where q is the number of parameters that are included in model 1, but not in model 2.

3. Calculate the difference $SS_{e2} - SS_{e1}$. This will be related to a χ^2 distribution with q degrees of freedom.

4. We can now test hypotheses of type $H_0: \beta_1 = \beta_2 =, ..., \beta_q = 0$ by the F test
$$F = \frac{(SS_{e2} - SS_{e1})/q}{SS_{e1}/(n-p-1)}$$
with $(q, n-p-1)$ degrees of freedom.

Example 2.2 *For the example data in Table 2.1, we can test $H_0: \beta_1 = \beta_2 = 0$ given that $\beta_3 \neq 0$. The full model gives $SS_{e1} = 105.17$ on 11 d.f. A simple linear regression that only contains the variable x_3 gives the following output:*

```
Analysis of Variance

Source          DF    SS        MS        F       P
Regression      1     939.93    939.93    38.47   0.000
Residual Error  13    317.67    24.44
Total           14    1257.60
```

We find that $SS_{e2} = 317.67$ on 13 d.f. The difference $SS_{e1} - SS_{e2} = 317.67 - 105.17 = 212.5$ has $13 - 11 = 2$ d.f. The hypothesis is tested as
$$F = \frac{212.5/2}{105.17/11} = 11.113$$
on (2, 11) df. The 5% limit is 3.98; the 1% limit is 7.20 and the 0.1% limit is 13.81. Thus, the hypothesis is rejected at the 1% level; a computer calculated p value is $p = 0.0023$.

2.9 Prediction

If the model is correct, it is possible to use the estimated model for prediction. As in simple linear regression, two types of prediction may be of interest:

1. We want to predict the average value of y, given a set of values of $x_1, x_2, ..., x_p$.

2. We want to predict the outcome for one single observation, given a set of values of $x_1, x_2, ..., x_p$.

In both cases, the predicted value is
$$\hat{y} = b_0 + b_1 x_1 + b_2 x_2 + ... + b_p x_p.$$

2 Multiple linear regression

It is possible (with computer assistance) to calculate the estimated variance of $\hat{y}$. Denote this estimated variance with $s_{\hat{y}}^2$. This can be used to give a confidence interval for the predicted mean value (case 1) as

$$\hat{y} \pm t_{(1-\alpha/2,\ n-p-1)} s_{\hat{y}}.$$

If we want a prediction interval for a single observation, we should add s_e^2 as an estimate of the variance around the fitted function. The interval is then

$$\hat{y} \pm t_{(1-\alpha/2,\ n-p-1)} \sqrt{s_e^2 + s_{\hat{y}}^2}.$$

Tests can be obtained using the same results.

It is possible to ask most regression programs for predicted values and prediction intervals for values of x that are not among the data. For example, suppose that for the data in Table 2.1 the year 1990 has given $x_1 = 54$, $x_2 = 23$ and $x_3 = 32$. What will be the expected quality? Minitab reports the results in the following form. CI is the confidence interval and PI is the prediction interval.

```
Predicted Values for New Observations

New Obs     Fit      SE Fit       95.0% CI            95.0% PI
1         79.633     1.408   ( 76.534;  82.732)   ( 72.155;  87.111)

Values of Predictors for New Observations

New Obs   Rain_W   Av_temp   Rain_H
1          54.0     23.0      32.0
```

2.10 Model building

The purpose of modeling is to give a simplified picture of some real phenomenon. A good model should contain most of the systematic variation in the data, such that the remaining variation could be regarded as "random". However, the model should not contain too much: a tmodel that is too complex does not help us to understand reality.

2.10.1 Partial and sequential tests

In regression analysis, there may be a choice between several variables: should the variables be included in the model or not? This choice is made more complicated by the fact that hypothesis tests in regression analysis depend on the order in which the variables were included in the model. As a study object, let us use three printouts of analyses from the data in Table 2.1: one analysis with only variable x_3, one analysis with only x_2 and x_3, and one analysis with all three variables. These results are as follows:

Regression Analysis: Quality versus Rain_H

```
The regression equation is
Quality = 91.9 - 0.260 Rain_H

Predictor        Coef      SE Coef        T        P
Constant        91.927       2.255      40.77    0.000
Rain_H         -0.26001     0.04192     -6.20    0.000

S = 4.943       R-Sq = 74.7%      R-Sq(adj) = 72.8%

Analysis of Variance

Source           DF        SS         MS        F        P
Regression        1      939.93     939.93    38.47    0.000
Residual Error   13      317.67      24.44
Total            14     1257.60
```

Regression Analysis: Quality versus Av_temp; Rain_H

```
The regression equation is
Quality = 48.6 + 1.66 Av_temp - 0.153 Rain_H

Predictor        Coef      SE Coef        T        P
Constant        48.61       11.88       4.09     0.001
Av_temp         1.6581      0.4504      3.68     0.003
Rain_H         -0.15323     0.04166    -3.68     0.003

S = 3.526       R-Sq = 88.1%      R-Sq(adj) = 86.2%

Analysis of Variance

Source           DF        SS         MS        F        P
Regression        2     1108.41     554.20    44.58    0.000
Residual Error   12      149.19      12.43
Total            14     1257.60
```

Regression Analysis: Quality versus Rain_W; Av_temp; Rain_H

```
The regression equation is
Quality = 48.9 + 0.0594 Rain_W + 1.36 Av_temp - 0.118 Rain_H

Predictor        Coef      SE Coef        T        P
Constant        48.91       10.41       4.70     0.001
Rain_W          0.05937     0.02767     2.15     0.055
Av_temp         1.3603      0.4187      3.25     0.008
Rain_H         -0.11773     0.04010    -2.94     0.014

S = 3.092       R-Sq = 91.6%      R-Sq(adj) = 89.4%

Analysis of Variance

Source           DF        SS         MS        F        P
Regression        3     1152.43     384.14    40.18    0.000
Residual Error   11      105.17       9.56
Total            14     1257.60
```

A summary of the results for the variable x_3 (Rain_H) is as follows:

2 Multiple linear regression

Analysis	Estimate of β_3	Std.error	t	p
1	-0.2600	0.04192	-6.20	0.000
2	-0.1532	0.04166	-3.68	0.003
3	-0.1177	0.04010	-2.94	0.014

We can see that the estimates as well as the standard errors of β_3 are different in the three analyses. The p values are also different. In fact, it is not unusual that a variable that is highly significant in one model may turn out to be "not significant" in another model.

The reason for this is that the x variables may be correlated. This means that the inclusion of one variable in the model may change the estimates of parameters for another variable. To make clear what we mean by a test or a calculation of sums of squares, we use the following terms:

Type 1: A partial sum of squares, or a partial test, for one parameter is based on the change in regression SS when that parameter is added to the model, in the order given in the MODEL statement. Thus, if we have variables x_1, x_2 and x_3 in that order, the sums of squares will be SS_{x_1}, $SS_{x_2|x_1}$ and $SS_{x_3|x_1,x_2}$.where, for example, $SS_{x_3|x_1,x_2}$ reads "the sum of square for x_3 when x_1 and x_2 are already in the model".

Type 2: A "type 2" test is based on the change in regression sum of squares when the variable in question is added last to the model. Thus, the sums of squares would be $SS_{x_1|x_2,x_3}$, $SS_{x_2|x_1,x_3}$ and $SS_{x_3|x_1,x_2}$. The t tests in multiple regression outputs are type 2-tests.

Later (Page 75) we will discuss tests of type 3 and type 4.

2.10.2 Stepwise regression

When there is a choice between many x variables, it is possible to ask the computer to choose among the variables in a systematic way. The following methods for automatic selection of variables is often present in regression software:

Backward elimination

First, all variables are included in the model. The variable that has the largest p-value is deleted. The model is re-run, and the process continues as long as all variables have p values smaller than a predetermined limit, for example $p = 0.15$.

Forward selection

The variable that has the strongest relation with y (i.e. the smallest p value) is included first. Then, the variable that has the strongest relation with y given the first variable is added, and so on. The process continues as long as all included variables have a p value smaller than a predetermined limit, for example $p = 0.15$.

Stepwise selection

This method works as forward selection, but in each step variables that are already in the model are tested; if the p value has increased above a certain limit, the variable may be deleted from the model.

Other methods such as maximum R^2 improvement or maximum R^2 selection may also be available in your software.

A warning is needed regarding stepwise methods, however. Such methods are not guaranteed to always reach the "best" model. To be certain that all possibilities have been checked, you would need to make all possible regressions and select the one that fulfills some criterion. A possible criterion is to choose the model that has the largest adjusted R^2. Another option is to use the one for which the Akaike information criterion, AIC is optimal; the AIC is further discussed on Page 202. If there are many x variables, it may not be feasible to make all possible regressions. In such cases, the stepwise methods may be used, but with caution.

2 Multiple linear regression

2.11 Exercises

Exercise 2.1 *The following variables have been recorded for seven farms:*

      Yield   Amount of yield of wheat per acre
      Fertil   Amount of fertilizer used per acre
      Rain   Amount of rainfall on the field

Data:

Yield	Fertil	Rain
40	100	10
50	200	20
50	300	10
70	400	30
65	500	20
65	600	20
80	700	30

Perform a multiple regression analysis (by computer) to study how yield depends on fertilizer and rainfall. Also, make diagnostic plots to check the assumptions.

Exercise 2.2 *Below is a set of data on the oil consumption for the heating of a number of greenhouses. For each observation we are given the following variables:*

   y Oil consumption for heating the greenhouse
   x_1 Average outdoor temperature
   x_2 Area of the greenhouse

y	x_1	x_2
140	17.8	170
200	16.6	210
370	12.2	150
600	7.1	190
620	2.8	110
1300	0.1	250
1050	−2.9	140
1280	−3.1	155
1100	−0.7	180
550	4.4	130

A. Make a graph of the relation between area and oil consumption.
B. Calculate a simple linear regression for the relation between area and oil consumption.

C. Calculate the correlation between area and oil consumption.

D. According to your calculations, is there any significant relation between area and oil consumption? Would you be prepared to announce that conclusion in a scientific paper?

E. Make a simple linear regression for the relation between temperature and oil consumption. Does the relation seem reasonable?

F. Make a multiple regression where y is related to both x_1 and x_2.

G. In the multiple regression, is there any significant relation between area and oil consumption?

H. How much does the average oil consumption change if you increase the area of a greenhouse by one m^2?

i) According to your analysis in B.

ii) According to your analysis in F. Which of the two do you consider more correct?

I. What oil consumption would be expected for a greenhouse with an area of 200 m^2 during a period with a temperature of $-1.5\ °C$?

J. Regardless of which regression model you use, there will still be some unexplained variation (residual variation). Parts of the residual variation are factors that have not been measured. Suggest a few factors that may have reduced the residual variance, if they had been measured and included in the model.

Exercise 2.3 *A researcher studied the relation between weight (pounds) and different variables for a sample of 40 males. The aim was to find a model that could predict the weight based on a set of x variables. Possible x variables were:*

x_1 = the height (inches) of the person
x_2 = waist circumference of the person (cm)
x_3 = cholesterol level of the person

The following models were studied.

variables	R^2	R^2_{Adj}	p	Equation
x_1, x_2, x_3	0.880	0.870	0.000	$-199 + 2.55x_1 + 2.18x_2 - 0.005x_3$
x_1, x_2	0.877	0.870	0.000	$-206 + 2.66x_1 + 2.15x_2$
x_1, x_3	0.277	0.238	0.002	$-148 + 4.65x_1 + 0.00589x_3$
x_2, x_3	0.804	0.793	0.000	$-42.8 + 2.41x_2 - 0.0106x_3$
x_1	0.273	0.254	0.001	$-139 + 4.55x_1$
x_2	0.790	0.785	0.000	$-44.1 + 2.37x_2$
x_3	0.001	0.000	0.874	$173 - 0.00233x_3$

Questions:

A. If only one predictor variable is used for predicting weight, which single variable is best? Why?

B. If exactly two predictor variables are used, which two variables should be chosen? Why?

2 Multiple linear regression

C. Which regression equation is best for predicting the weight? Why?

D. If a male has a height of 72 inches, a waist circumference of 105 cm and a cholesterol level of 250 mg, what is the best predicted value of his height?

Chapter 3

Nonlinear models

3.1 Models that are linear in the parameters

Multiple regression is "linear", in the sense that the parameters (β_0, β_1 etc.) enter linearly into the models. This means that the parameters are not included in nonlinear expressions. However, it is quite possible to include nonlinear functions of x among the independent variables.

A simple example of this is polynomial regression. A general polynomial regression model of degree p can be written as

$$y_i = \beta_0 + \beta_1 x_i + \beta_2 x_i^2 + \beta_3 x_i^3 + \ldots + \beta_p x_i^p + e_i.$$

This type of model can be fitted using standard multiple regression software. All we need to do is to ask our statistics program to calculate the values of x^2, x^3 etc. and store these as new variables in the data. This can be illustrated with an example:

Example 3.1 *(Polynomial model).* Table 3.1 displays a set of data from Snedecor and Cochran (1980). The purpose of the experiment was to study the relation between protein content (y) and yield (x) of wheat. A plot of the data reveals that the relationship seems nonlinear. A simple model that may account for the nonlinearity is

$$y_i = \beta_0 + \beta_1 x_i + \beta_2 x_i^2 + e_i$$

which is a second degree polynomial model.

Our analysis of these data uses the SAS procedure REG. The data are

3 Nonlinear models

Table 3.1: Relation between protein content (y) and yield (x) of wheat.

x	y
5	16.2
8	14.2
10	14.6
11	18.3
14	13.2
16	13.0
17	13.0
17	13.4
18	10.6
20	12.8
22	12.6
24	11.6
26	11.0
30	9.8
32	10.4
34	10.9
36	12.2
38	9.8
43	10.7

read into a dataset, and SAS is asked to create a new variable X2 which contains the squares of the x values. Then, a multiple linear regression of y on x and x^2 is performed. The program is as follows:

```
DATA poly;
   INPUT x y;
   x2=x*x;
   /*This asks SAS to calculate x-squared and names this variable x2*/
CARDS;
5       16.2
8       14.2
10      14.6
... (Data lines omitted)
36      12.2
38      9.8
43      10.7
;
PROC REG data=poly;
   MODEL y = x x2;
   OUTPUT out=ut P=pred R=res;
RUN;
```

The following output was obtained:

Models that are linear in the parameters

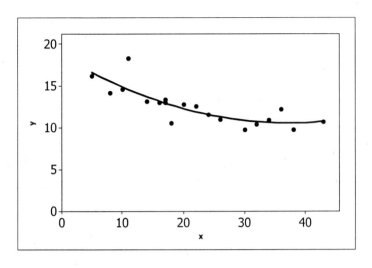

Figure 3.1: A quadratic regression model fitted to the data in Table 3.1.

```
                    The REG Procedure
                      Model: MODEL1
                    Dependent Variable: y

                    Analysis of Variance

                            Sum of           Mean
Source              DF     Squares         Square    F Value    Pr > F

Model                2    61.73438       30.86719      18.77    <.0001
Error               16    26.31193        1.64450
Corrected Total     18    88.04632


            Root MSE              1.28238    R-Square     0.7012
            Dependent Mean       12.54211    Adj R-Sq     0.6638
            Coeff Var            10.22459
                     Parameter Estimates

                    Parameter     Standard
Variable     DF     Estimate        Error    t Value    Pr > |t|

Intercept     1     18.67672      1.37210      13.61    <.0001
x             1     -0.43673      0.12936      -3.38     0.0039
x2            1      0.00587      0.00266       2.20     0.0427
```

We find that the estimated equation is

$$\hat{y} = 18.6767 - 0.4367x + 0.0059x^2.$$

A graph of the data, along with the fitted curve, is given in Figure 3.1.

Polynomial models are often useful, since most functions can be approximated by polynomials. It should be noted that the degree of the polynomial is limited by the number of distinct x values. If the data con-

3 Nonlinear models

tain p distinct x values, then we can fit a polynomial of at most degree $p-1$.

It should be pointed out that other functions than polynomials can also be used. For example, periodic events can be modeled using functions such as cos, sin etc. Thus, a model such as

$$y = \beta_0 + \beta_1 \sin(x) + e$$

can be estimated by including $\sin(x)$ as a new variable and using a linear regression program.

3.2 Models that can be linearized by transformation

Some types of models that are not linear in the parameters can be made linear by transformation. If the required regression equation is

$$E(y) = f(x)$$

where $f(x)$ is nonlinear, the principle is to find some transformation $g(y)$ that is such that $g(f(x))$ is linear. This often works fine. A small warning should be issued, however. If the original model is $y = f(x) + e$, where the residuals are normal and homoscedastic, then the transformation would change the behavior of the residuals. When y is transformed, we have to assume that the residuals *after* transformation are normal and homoscedastic. This assumption can be examined using standard residual diagnostics.

Transformations are used for two main reasons. One reason is to "linearize" a regression model. A second reason may be to change the behavior of the residuals, so that they are more nearly normal and homoscedastic. In both cases, residual diagnostics on the estimated residuals should be used to assess the effects of the transformation.

Example 3.2 *(Exponential model).* Gowen and Price counted the number of lesions of aucuba mosaic virus after exposure to X-rays for

Models that can be linearized by transformation

various times. The results were:

Minutes exposure	Count (hundreds)	log(count)
0	271	5.602
15	108	4.682
30	59	4.078
45	29	3.367
60	12	2.485

It was assumed that the count (y) depends on the exposure time (x) through an exponential relation of type

$$y = a \cdot e^{-bx}.$$

A convenient way to estimate the parameters of such a function is to calculate the logarithm of each side of the equation; in this case, we used log to the base e. We then get

$$\log y = \log a - bx$$
$$y' = a' - bx.$$

The parameters of this model can be estimated by making a linear regression of $y' = \log(y)$ on x. We feed the data into Minitab, and ask Minitab ("Calc") to calculate the log of y, as indicated in the table above. A linear regression of $\log(y)$ on x gives the following output:

Regression Analysis: log(count) versus Exposure

```
The regression equation is
log(count) = 5.55 - 0.0503 Exposure

Predictor      Coef      SE Coef        T        P
Constant     5.55265     0.07160      77.55    0.000
Exposure    -0.050328    0.001949    -25.83    0.000

S = 0.09243     R-Sq = 99.6%     R-Sq(adj) = 99.4%

Analysis of Variance

Source           DF        SS         MS        F         P
Regression        1      5.6991     5.6991    667.04    0.000
Residual Error    3      0.0256     0.0085
Total             4      5.7248
```

The model is estimated as

$$\widehat{\log(y)} = 5.55 - 0.0503x.$$

3 Nonlinear models

We can exponentiate both sides of the expression to obtain

$$\widehat{y} = e^{5.55} e^{-0.0503x}$$
$$= 257.24 \cdot e^{-0.0503x}$$

We can ask Minitab to calculate the fitted values of $\log(y)$ and exponentiate these. Then, the data and the fitted model can be plotted in the same graph. Of course, it is also possible to make a graph of the logged model. These two graphs are given in Figure 3.2 and Figure 3.3. The data show a good fit to the model.

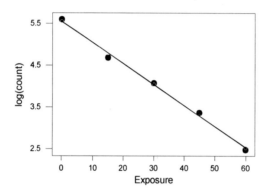

Figure 3.2: Relation between y and x: logarithmic scale.

3.3 Box-Cox transformations

Box and Cox (1964) suggested a useful family of transformations, defined by

$$h(y; \lambda) = \begin{cases} (y^\lambda - 1)/\lambda & \text{for } \lambda \neq 0 \\ \log(y) & \text{for } \lambda = 0 \end{cases}$$

The transformation parameter λ in the Box-Cox family of transformations is often suggested by earlier experience with the kind of data in question. Alternatively, it can be suggested by the data. Note that it is the distribution of the residuals, not the distribution of the raw data y, that should be near normal and homoscedastic.

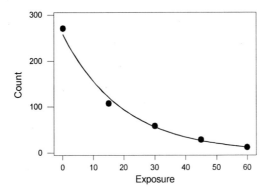

Figure 3.3: Relation between y and x: original scale.

The Minitab package includes a routine for Box-Cox transformations as a part of the quality control section. The routine suggests a value of λ based on your data, and presents a graph that can be interpreted as a kind of confidence interval for λ. It is common practice to round the value of λ such that, for example, a value close to 0 would give a log transformation, and so on according to Table 3.2.

Table 3.2: Examples of Box-Cox transformations.

Value of λ	Transformation
2	$y' = y^2$
1	$y' = y$
0.5	$y' = \sqrt{y}$
0	$y' = \log(y)$
-0.5	$y' = \frac{1}{\sqrt{y}}$
-1	$y' = \frac{1}{y}$

3.4 Truly nonlinear models

Some types of models cannot be transformed into a form that can be handled by software for linear regression. An example is the so called

3 Nonlinear models

logistic growth function

$$y = \frac{\beta_1}{1 + e^{-(x-\beta_2)/\beta_3}}.$$

This type of model is sometimes used to model how the size of e.g. animals depends on age (x). As age tends towards infinity, the size tends towards an asymptote, β_1. This type of "intrinsically nonlinear models" requires special software, like the SAS procedure NLIN.

As another example, consider the model

$$y = \beta_0 + e^{\beta_1 x} + e.$$

Given a set of data, we would estimate the model parameters by minimizing

$$S = SS_e = \sum \left(y_i - b_0 - e^{b_1 x_i}\right)^2.$$

The derivatives of SS_e with respect to b_0 and b_1 are

$$\frac{\partial S}{\partial b_0} = -2 \sum \left(y_i - b_0 - e^{b_1 x_i}\right)$$

$$\frac{\partial S}{\partial b_1} = -2 \sum \left(y_i - b_0 - e^{b_1 x_i}\right) \cdot e^{b_1 x_i} \cdot x_i$$

We should set these derivatives equal to zero and solve for b_0 and b_1. In this case, this is not too difficult, but in general this type of problems may be solved using the computer.

Example 3.3 *Suppose that the growth of a type of bacteria for theoretical reasons is assumed to follow the function $y = \beta_0 + e^{\beta_1 x} + e$ where x is time. Eleven colonies have been allowed to grow, under laboratory* conditions, with the following results:

x	y
0	1.465
5	4.644
10	8.988
15	16.585
20	10.999
25	16.201
30	26.052
35	38.157
40	55.553
45	95.735
50	157.923

The following SAS program is used to enter the data and fit the model:

```
DATA growth;
INPUT x    y;
CARDS;
0     1.465
5     4.644
10    8.988
15    16.585
20    10.999
25    16.201
30    26.052
35    38.157
40    55.553
45    95.735
50    157.923
;
PROC NLIN data=growth;
MODEL y = b0 + exp(b1*x);
PARMS b0=1 b1=1; /*Initital guesses of parameters*/
OUTPUT out=ut p=pred r=res;
RUN;
```

Very rough initial guesses of the parameters were used. It is possible, but not necessary, to present formulas for the derivatives to the program. In this case the derivatives were not included, so the program will calculate these numerically. The following output was obtained:

3 Nonlinear models

```
                    Estimation Summary

             Method               Gauss-Newton
             Iterations                     49
             R                        4.01E-7
             PPC(b0)                 1.792E-7
             RPC(b0)                 0.000054
             Object                   7.553E-8
             Objective                105.3423
             Observations Read              11
             Observations Used              11
             Observations Missing            0
                    Sum of      Mean                 Approx
Source         DF   Squares     Square    F Value    Pr > F

Regression      2   39983.3    19991.7    1964.49    <.0001
Residual        9     105.3    11.7047
Uncorrected Total 11  40088.7


Corrected Total 10   23099.1
                      Approx      Approximate 95% Confidence
Parameter      Estimate  Std Error            Limits

b0             4.4683    1.2295       1.6869      7.2497
b1             0.1005    0.000451     0.0994      0.1015
                Approximate Correlation Matrix
                          b0                b1

         b0        1.0000000         -0.5441624
         b1       -0.5441624          1.0000000
```

Estimates of the parameters are given along with estimates of their standard errors and correlations. The estimated regression function is

$$\widehat{y} = 4.4683 + e^{0.1005x}.$$

A plot of the relation between y and x, including the fitted function, is presented in Figure 3.4

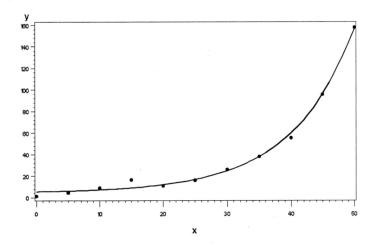

Figure 3.4: Relation between y and x in the bacterial growth example.

3.5 "Transform both sides"

Suppose that you want to fit the nonlinear model

$$y = f(x) + e$$

to a set of data. Initial analyses suggest that the residuals e do not seem to be normal and homoscedastic. One alternative is then to transform y (for example, using a Box-Cox transformation; see page 58). This would give a new model

$$h(y) = f^*(x) + e^*$$

where the new residuals e^* may be closer to normality and homoscedasticity.

However, the model is no longer the model that was suggested by theory. For example, if $f(\cdot)$ is the linear model $y = \alpha + \beta x$, we can no longer expect the relation between $h(y)$ and x to be linear. Our transformation has, in a sense, destroyed the functional relationship $f(\cdot)$ between y and x, and we have to choose some other function $f^*(\cdot)$ for our model fitting.

In some model-fitting situations this is not a serious problem, but in situations where the function $f(\cdot)$ was suggested by some theory, or where the parameters of $f(\cdot)$ have some meaningful interpretation, it is

3 Nonlinear models

unsatisfactory to have to step back from the original model.

As a solution to this problem, Carroll & Ruppert (1984; 1988) suggested the following approach. Suppose that analysis of the data suggests that some monotonic transformation $h(y)$ of the dependent variable is needed to give residuals that are more nearly normal and homoscedastic. Then, the model fitting would amount to fitting the model

$$h(y) = h(f(x)) + e$$

using some (generally) nonlinear regression method. Thus, *both* sides of the regression equation would be transformed using the transformation $h(\cdot)$.

Parameter estimates obtained by fitting this model to the data are biased estimates of the parameters of the original model. However, Carrol & Ruppert state that this bias is generally negligible.

For the case of Box-Cox transformations suggested by the data, Carroll & Ruppert discuss possible effects of regarding the transformation parameter λ as fixed. They state that, at least asymptotically, the limiting distribution of the parameter estimators is the same whether λ is known or unknown, and that the relative efficiency (compared to a model with λ known) is at least $2/\pi$; simulations indicate that the practical efficiency is often even close to 1. The tentative practical conclusion would seem to be that estimates, standard errors etc. obtained from fitting the transformed model to data, using some transformation $h(\cdot)$, can be used for inference.

Example 3.4 *Didon and Olsson (1997) described a bioassay method for the detection of two different herbicides. Seedlings were grown in petri dishes that were exposed to different herbicide concentrations. The average root growth of the seedlings in each petri dish during 24h was measured. Root growth was measured as a percentage of untreated controls. Root growth was somewhat stimulated at low herbicide levels, but for higher doses, the root growth was retarded. For theoretical reasons, the three-parameter logistic dose-response model*

$$y = \frac{D}{1 + e^{(b \log x - g)}} + e$$

was chosen as a reasonable representation of the data. The asymptote D was allowed to differ from 100% because of the stimulation at low doses, mentioned above. Preliminary analysis of variance showed that the variance was larger for small doses, i.e. when the average root growth was large. The search for a good Box-Cox transformation indicated a

"Transform both sides"

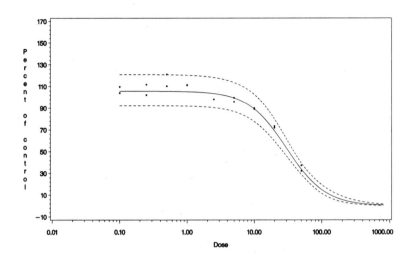

Figure 3.5: Dose-response curve and 95% prediction limits for the root growth of Brassica rapa subjected to MCPA, as a percentage of control. Observations are mean values of 10 petri dishes (70 seedlings). Continuous line denotes the fitted function, dashed lines are upper and lower 95% limits.

value of λ close to 0; thus, a logarithmic transformation was used for the analyses.

The procedure NLIN in the SAS (2008) package was used to fit the model

$$\log(y) = \log\left(\frac{D}{1+e^{(b\log x - g)}}\right) + e$$

to the data for each herbicide. This was done both for data from individual petri dishes, and for the average of groups of 10 petri dishes. 95% limits for individual observations were computed. These prediction limits were used in inverse prediction of the dose for given values of y. Only the results for the average of 10 petri dishes for one of the herbicides (MCPA) are presented here. For a complete account, see Didon and Olsson (1997).

3 Nonlinear models

3.6 Exercises

Exercise 3.1 *An experiment was performed in 1987 on 89 randomly selected rats, aged between 0 and 20 days. The purpose of the experiment was to estimate the relationship between oxygen uptake (VO2 ; Unit: ml/h) and age. The data were fitted to the following equation:*

$$\begin{array}{rcl} log(VO2) & = & log(1.715) \;\; + \;\; 1.004 \; log(age). \\ & & (0.0134) \qquad\qquad (0.0182) \end{array}$$

(Standard errors of the coefficients are given within parenthesis. Logs were taken to base 10).

A hypothesis test indicated that the slope was not significantly different from one. In 1956, another researcher had suggested that "One unit increase in log(age), on the average, gives 0.75 units increase in log(VO2)".

A. Is there enough evidence in these data to reject the suggestion made in 1956?

B. The correlation coefficient for the logged data was 0.986. Test if the true (population) correlation is larger than zero. State your assumptions, and give a verbal conclusion.

C. For the 89 rats in the experiment the mean of the logged age was 0.83, and $\sum(x-\bar{x})^2 = 56.2877$, where $x = log(age)$. The estimated residual variance, $s_e^2 = 0.01$. Use this information to predict the logged oxygen uptake of a single rat aged 15 days, and give a confidence interval for your prediction. Then translate your interval into the original unit (VO2).

Exercise 3.2 *In a study of the growth of chicken embryos, the dry weight of a number of chicken embryos was measured. The embryos were of different ages. For theoretical reasons, it was assumed that an exponential growth model of the form $y = \alpha \cdot \beta^x$ would describe the relation in an appropriate way. The following data were available:*

Age (weeks)	Dry weight
x	y
6	0.029
8	0.079
10	0.181
12	0.425
14	1.130
16	2.812

Exercises

A. Estimate the parameters α and β of the model from the data given above.

B. What assumptions about the residuals are needed?

C. What weight y do you expect in a chicken embryo of age 11 weeks?

Chapter 4

General linear models

4.1 The role of models

Many of the methods taught during elementary statistics courses can be collected under the heading of *general linear models*, GLM. Statistical packages like SAS, Minitab and R have standard procedures for general linear models. GLMs include regression analysis, analysis of variance, and analysis of covariance. Some applied researchers are not aware that even their simplest analyses are, in fact, model based.

Models play an important role in statistical inference. A model is a mathematical way of describing the relationships between a response

4 General linear models

variable and a set of independent variables. Some models can be seen as a theory about how the data were generated. Other models are only intended to provide a convenient summary of the data. Statistical models, as opposed to deterministic models, account for the possibility that the relationship is not perfect. This is done by allowing for unexplained variation, in the form of residuals.

A way of describing a frequently used class of statistical models is

$$\text{response} = \text{systematic component} + \text{residual component}$$

Models of this type are, at best, approximations of the actual conditions. A model is seldom "true" in any real sense. The best we can look for may be a model that can provide a reasonable approximation to reality. However, some models are certainly better than others. The role of the researcher is to find a model that is reasonable, while at the same time it is simple enough to be interpretable.

4.2 General linear models

In a general linear model (GLM), the observed value of the dependent variable y for observation number i ($i = 1, 2, ..., n$) is modeled as a linear function of p so-called independent variables $x_1, x_2, \ldots, x_p$ as

$$y_i = \beta_0 + \beta_1 x_{i1} + \ldots + \beta_p x_{ip} + e_i.$$

This, in essence, is a kind of multiple regression model. In this chapter we will show that this kind of model can be used in a more general context than for regression analysis only. However, some of the following results are the same as in the multiple regression chapter.

The parameters of the model are $\beta_0, \beta_1, \ldots, \beta_p$, and e_i is a residual. It is common to assume that the residuals are independent, normally distributed and that the variances are the same for all e_i. Some models do not contain any intercept term β_0.

The purpose of the analysis may be model building, estimation, prediction, hypothesis testing, or a combination of these. We will briefly summarize some results on estimation and hypothesis testing in general linear models. For a more complete description, reference is made to standard textbooks in regression analysis, such as Draper and Smith (1998) or Sen and Srivastava (1990); and textbooks in analysis of variance, such as Montgomery (1984) or Christensen (1996).

4.3 Estimation

Estimation of parameters in general linear models is often done using the method of least squares. For normal theory models this is equivalent to maximum likelihood estimation. The parameters are estimated with the values for which the sum of the squared residuals, $\sum_i e_i^2$, is minimal.

Thus, we should minimize

$$S = \sum_i e_i^2 = \sum_i \left(y_i - (\beta_0 + \beta_1 x_{i1} + \ldots + \beta_p x_{ip})\right)^2$$

as a function of $\beta_0, \beta_1, \ldots, \beta_p$. This yields an equation system with $(p+1)$ equations and $(p+1)$ unknowns, namely $\beta_0, \beta_1, \ldots, \beta_p$. Solving this equation system, the so-called normal equations, is best left to a computer program. The result is estimated values, $\widehat{\beta}_0, \widehat{\beta}_1, \ldots, \widehat{\beta}_p$. Throughout this text we will use a "hat", $\widehat{}$, to symbolize an estimator.

It may happen that the equation system does not have a unique solution. In such cases, the computer program may use a so-called generalized inverse and suggest one of many solutions. Alternatively we can restrict the number of parameters in the model by introducing constraints that lead to a unique solution.

4.4 Assessing the fit of the model

4.4.1 Predicted values and residuals

When the parameters of a general linear model have been estimated, you may want to assess how well the model fits the data. This is done by subdividing the variation in the data into two parts: systematic variation and unexplained variation. Formally, this is done as follows.

We define the predicted value (or fitted value) of the response variable as

$$\begin{aligned}\widehat{y}_i &= \widehat{\beta}_0 + \widehat{\beta}_1 x_{i1} + \widehat{\beta}_2 x_{i2} + \ldots + \widehat{\beta}_p x_{ip} \\ &= \sum_{j=0}^{p} \widehat{\beta}_j x_{ij}.\end{aligned}$$

The predicted values are the values that we would get on the dependent variable if the model had been perfect, i.e. if all residuals had been zero. The difference between the observed value and the predicted value is the observed residual:

$$\widehat{e}_i = y_i - \widehat{y}_i.$$

4.4.2 Sums of squares decomposition

The total variation in the data can be measured as the total sum of squares,
$$SS_T = \sum_i (y_i - \overline{y})^2.$$

This can be subdivided as

$$\begin{aligned}\sum_i (y_i - \overline{y})^2 &= \sum_i (y_i - \widehat{y}_i + \widehat{y}_i - \overline{y})^2 \\ &= \sum_i (y_i - \widehat{y}_i)^2 + \sum_i (\widehat{y}_i - \overline{y})^2 + 2\sum_i (y_i - \widehat{y}_i)(\widehat{y}_i - \overline{y}).\end{aligned}$$

The last term can be shown to be zero. Thus, the total sum of squares SS_T can be subdivided into two parts:

$$SS_{Model} = \sum_i (\widehat{y}_i - \overline{y})^2$$

and

$$SS_e = \sum_i (y_i - \widehat{y}_i)^2.$$

SS_e, called the residual (or error) sum of squares, will be small if the model fits the data well.

It holds that

$$SS_T = \sum_i (y_i - \overline{y})^2 \text{ has } n-1 \text{ degrees of freedom } (d.f.).$$

$$SS_{Model} = \sum_i (\widehat{y}_i - \overline{y})^2 \text{ has } p \text{ d.f.}$$

$$SS_e = \sum_i (y_i - \widehat{y}_i)^2 \text{ has } n-p-1 \text{ d.f.}$$

The subdivision of the total variation (the total sum of squares) into parts is often summarized as an analysis of variance table:

Source	Sum of squares (SS)	d.f.	$MS = SS/d.f.$
Model	SS_{Model}	p	MS_{Model}
Residual	SS_e	$n-p-1$	$MS_e = \widehat{\sigma}^2$
Total	SS_T	$n-1$	

These results can be used in several ways. MS_e provides an estimator of σ^2, which is the variance of the residuals. A descriptive measure of the fit of the model to data can be calculated as

$$R^2 = \frac{SS_{Model}}{SS_T} = 1 - \frac{SS_e}{SS_T}.$$

R^2 is called the coefficient of determination. It holds that $0 \leq R^2 \leq 1$. For data where the predicted values $\hat{y}_i$ are all equal to the corresponding observed values y_i, R^2 would be 1. It is not possible to judge a model based on R^2 alone. In some applications, for example econometric model building, models often have values of R^2 very close to 1. In other applications, models can be valuable and interpretable also when R^2 is rather small. When several models have been fitted to the same data, R^2 can be used to judge which model to prefer. However, since R^2 increases (or is unchanged) when new terms are added to the model, model comparisons are often based on the adjusted R^2. The adjusted R^2 decreases when irrelevant terms are added to the model. It is defined as

$$R^2_{adj} = 1 - \frac{n-1}{n-p-1}\left(1 - R^2\right) = 1 - \frac{MS_e}{SS_T/(n-1)}.$$

This can be interpreted as

$$R^2_{adj} = 1 - \frac{\text{Variance estimated from the model}}{\text{Variance estimated without any model}}.$$

A formal test of the full model (i.e. a test of the hypothesis that $\beta_1, ..., \beta_p$ are all zero) can be obtained as

$$F = \frac{MS_{Model}}{MS_e}.$$

This is compared to appropriate percentage points of the F distribution with $(p, n-p-1)$ degrees of freedom.

4.5 Inference on single parameters

Parameter estimators in general linear models are linear functions of the observed data. Thus, the estimator of any parameter β_j can be written as

$$\widehat{\beta}_j = \sum_i w_{ij} y_i$$

where w_{ij} are known weights. If we assume that all y_i:s have the same variance σ^2, this makes it possible to obtain the variance of any para-

4 General linear models

meter estimator as
$$Var\left(\widehat{\beta}_j\right) = \sum_i w_{ij}^2 \sigma^2.$$

The variance σ^2 can be estimated from data as
$$\widehat{\sigma}^2 = \frac{\sum_i \widehat{e}_i^2}{n-p-1} = MS_e.$$

The variance of a parameter estimator $\widehat{\beta}_j$ can now be estimated as
$$\widehat{Var}\left(\widehat{\beta}_j\right) = \sum_i w_{ij}^2 \widehat{\sigma}^2.$$

This makes it possible to calculate confidence intervals and to test hypotheses about single parameters. A test of the hypothesis that the parameter β_j is zero can be made by comparing
$$t = \frac{\widehat{\beta}_j}{\sqrt{\widehat{Var}\left(\widehat{\beta}_j\right)}}$$

with the appropriate percentage point of the t distribution with $n-p-1$ degrees of freedom. Similarly,
$$\widehat{\beta}_j \pm t_{(1-\alpha/2, n-p-1)} \sqrt{\widehat{Var}\left(\widehat{\beta}_j\right)}$$

would provide a $(1-\alpha) \cdot 100\%$ confidence interval for the parameter β_j.

4.6 Tests on subsets of the parameters

In some cases it is of interest to make simultaneous inference about several parameters. For example, in a model with p parameters one may wish to simultaneously test if q of the parameters are zero. This can be done in the following way:

1. Estimate the parameters of the full model. This will give an error sum of squares, SS_{e1}, with $(n-p-1)$ degrees of freedom.

2. Estimate the parameters of the smaller model, i.e. the model with fewer parameters. This will give an error sum of squares, SS_{e2}, with $(n-p+q-1)$ degrees of freedom, where q is the number of parameters that are included in model 1, but not in model 2.

3. The difference $SS_{e2} - SS_{e1}$ will be related to a χ^2 distribution with q degrees of freedom. We can now test hypotheses of type H$_0$: $\beta_1 = \beta_2 =, ..., \beta_q = 0$ by the F test

$$F = \frac{(SS_{e2} - SS_{e1})/q}{SS_{e1}/(n-p-1)}$$

with $(q, n-p-1)$ degrees of freedom.

4.7 Different types of tests

Tests of single parameters in general linear models depend on the order in which the hypotheses are tested. Tests in balanced analysis of variance designs are exceptions; in such models the different parameter estimates are independent. In other cases there are several ways to test hypotheses. SAS handles this problem by allowing the user to select among four different types of tests.

Type 1 means that the test for each parameter is calculated as the change in SS_e when the parameter is added to the model, in the order given in the MODEL statement. If we have the model Y = A B A*B, SS_A is calculated first as if the experiment had been a one-factor experiment (model: Y=A). Then $SS_{B|A}$ is calculated as the reduction in SS_e when we run the model Y=A B, and finally the interaction $SS_{AB|A,B}$ is obtained as the reduction in SS_e when we also add the interaction to the model. This can be written as $SS(A)$, $SS(B|A)$ and $SS(AB|A,B)$. Type I SS are sometimes called sequential sums of squares.

Type 2 means that the SS for each parameter is calculated as if the factor had been added last to the model except that, for interactions, all main effects that are part of the interaction should also be included. For the model Y = A B A*B this gives the SS as $SS(A|B)$; $SS(B|A)$ and $SS(AB|A,B)$.

Type 3 is, loosely speaking, an attempt to calculate what the SS would have been if the experiment had been balanced. These are often called partial sums of squares. These SS cannot in general be computed by comparing model SS from several models. The Type-3 SS are generally preferred when experiments are unbalanced. One problem with them is that the sum of the SS for all factors and interactions is generally not the same as the total SS. Minitab gives the Type 3 SS as "Adjusted Sum of Squares".

4 General linear models

Type 4 differs from type 3 in the method of handling empty cells, i.e. incomplete experiments.

If the experiment is balanced, all these SS will be equal. In practice, tests in unbalanced situations are often done using type-3 SS (or "Adjusted Sum of Squares" in Minitab). Unfortunately, this is not an infallible method.

4.8 Some applications

4.8.1 Simple linear regression

In regression analysis, the data contain one column of values of y and another column of x values. The small regression model $y_i = \beta_0 + \beta_1 x_i + e_i$ with $n = 4$ observations can be written as

$$y_1 = \beta_0 + \beta_1 x_1 + e_1$$
$$y_2 = \beta_0 + \beta_1 x_2 + e_2$$
$$y_3 = \beta_0 + \beta_1 x_3 + e_3$$
$$y_4 = \beta_0 + \beta_1 x_4 + e_4$$

Example 4.1 *Heggestad and Bennet (1981) studied the effects of air pollution on the yield of Blue Lake snap beans. The beans were grown in field chambers that were fumigated with various levels of sulphur dioxide. After a month of fumigation, the plants were harvested and the yield was recorded. The results are reproduced in Table 4.1 and Figure 4.1.*

Table 4.1: Yield of Lake Snap beans following fumigation with SO_2.

x SO_2	y Yield
0.00	1.15
0.00	1.30
0.06	1.19
0.06	1.64
0.12	1.21
0.12	1.00
0.30	0.65
0.30	0.76

Figure 4.1 suggests that a linear trend may provide a reasonable approximation to the data, over the levels of SO_2 included in the experiment. The linear function fitted to these data is $\widehat{y} = 1.363 - 2.083x$.

Some applications

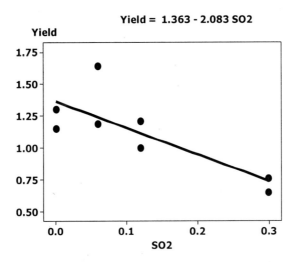

Figure 4.1: Relation between amount of sulphur dioxide and yield of Blue Lake snap beans. Raw data and fitted regression line.

A SAS regression output, including ANOVA table, is given below. It can be concluded that the yield decreases significantly with increasing SO2 level ($p = 0.0168$), the rate of decrease being about 2.08 units per unit of SO_2.

```
                       Analysis of Variance

                            Sum of          Mean
Source             DF      Squares        Square    F Value    Pr > F
Model               1      0.43750       0.43750      10.77    0.0168
Error               6      0.24365       0.04061
Corrected Total     7      0.68115

          Root MSE              0.20152    R-Square    0.6423
          Dependent Mean        1.11250    Adj R-Sq    0.5827
          Coeff Var            18.11372

                       Parameter Estimates

                     Parameter     Standard
Variable      DF      Estimate        Error    t Value    Pr > |t|
Intercept      1       1.36250      0.10429      13.06      <.0001
SO2            1      -2.08333      0.63471      -3.28      0.0168
```

4.8.2 Multiple regression

Generalization of simple linear regression models to include more than one independent variable is rather straightforward. For example, suppose that y may depend on two variables, and that we have made $n = 6$

4 General linear models

Table 4.2: Relation between the strength of wooden beams and the specific gravity and moisture content of the beam.

Strength	Gravity	Moisture
11.14	0.499	11.1
12.74	0.558	8.9
13.13	0.604	8.8
11.51	0.441	8.9
12.38	0.550	8.8
12.60	0.528	9.9
11.13	0.418	10.7
11.70	0.480	10.5
11.02	0.406	10.5
11.41	0.467	10.7

observations. The regression model is then $y_i = \beta_0 + \beta_1 x_{i1} + \beta_2 x_{i2} + e_i$, $i = 1, \ldots, 6$. If we write out the model for each observation we obtain

$$\begin{aligned}
y_1 &= \beta_0 + \beta_1 x_{11} + \beta_2 x_{12} + e_1 \\
y_2 &= \beta_0 + \beta_1 x_{21} + \beta_2 x_{22} + e_2 \\
y_3 &= \beta_0 + \beta_1 x_{31} + \beta_2 x_{32} + e_3 \\
y_4 &= \beta_0 + \beta_1 x_{41} + \beta_2 x_{42} + e_4 \\
y_5 &= \beta_0 + \beta_1 x_{51} + \beta_2 x_{52} + e_5 \\
y_6 &= \beta_0 + \beta_1 x_{61} + \beta_2 x_{62} + e_6
\end{aligned}$$

Example 4.2 *Some factors that may affect the strength of wood beams were studied by Draper and Stonemann (1966). The variables studied were:*

- *Strength = the strength of the wood beam when tested.*
- *Gravity = specific gravity of the beam.*
- *Moisture = the moisture content of the beam.*

A set of data of this type is reproduced in Table 4.2.

A multiple regression output from Minitab based on these data is as follows:

Regression Analysis: Strength versus Gravity; Moisture

```
The regression equation is
Strength = 10.3 + 8.49 Gravity - 0.266 Moisture

Predictor      Coef    SE Coef      T        P
Constant      10.302     1.896     5.43    0.001
Gravity        8.495     1.785     4.76    0.002
Moisture      -0.2663    0.1237   -2.15    0.068

S = 0.275440      R-Sq = 90.0%      R-Sq(adj) = 87.1%



Analysis of Variance

Source            DF      SS       MS       F        P
Regression         2    4.7792   2.3896   31.50    0.000
Residual Error     7    0.5311   0.0759
Total              9    5.3102
```

The output indicates that the predictor variables do have a relationship with the beam strength. The size and direction of this relationship is given by the estimated coefficients of the regression equation. It appears that the strength increases when specific gravity increases, and that it decreases with increasing moisture content.

4.8.3 t tests and dummy variables

Classification variables (non-numeric variables, "class variables"), such as treatments, groups or blocks can be included in the model as so called dummy variables, i.e. as variables that only take on the values 0 or 1. For example, a simple t test on data with two groups and three observations per group can be formulated as

$$y_{ij} = \mu + \beta d_i + e_{ij}$$
$$i = 1, 2, 3; \ j = 1, 2.$$

Here, μ is a general mean value, d_i is a dummy variable that has value $d_i = 1$ if observation i belongs to group 1 and $d_i = 0$ if it belongs to group 2, and e_{ij} is a residual. According to this model, the population mean value for group 1 is $\mu_1 = \mu + \beta$ and the population mean value for group 2 is simply $\mu_2 = \mu$. In the t-test situation we want to examine whether μ_1 is different from μ_2, i.e. whether β is different from 0. A simple model of this type, where there are two treatments, each with 3

4 General linear models

Table 4.3: Dopamine levels in the brains of rats under two treatments.

Dopamine, ng/kg	
Toluene group	Control group
3.420	1.820
2.314	1.843
1.911	1.397
2.464	1.803
2.781	2.539
2.803	1.990

observations, can be written as

$$y_{11} = \mu + \beta \cdot 1 + e_{11}$$
$$y_{21} = \mu + \beta \cdot 1 + e_{21}$$
$$y_{31} = \mu + \beta \cdot 1 + e_{31}$$
$$y_{12} = \mu + \beta \cdot 0 + e_{12}$$
$$y_{22} = \mu + \beta \cdot 0 + e_{22}$$
$$y_{32} = \mu + \beta \cdot 0 + e_{32}$$

Example 4.3 *In a pharmacological study (Rea et al, 1984), researchers measured the concentration of dopamine in the brains of six control rats and of six rats that had been exposed to toluene. The concentrations in the striatum region of the brain are given in Table 4.3.*

The interest lies in comparing the two groups with respect to average dopamine level. This is often done as a two sample t test. To illustrate that the t test is actually a special case of a general linear model, we analyzed these data with Minitab using regression analysis with group as a dummy variable. Rats in the toluene group were given the value 1 on the dummy variable, while rats in the control group were coded as 0. The Minitab output of the regression analysis is:

Regression Analysis

```
The regression equation is
Dopamine level = 1.90 + 0.717 Group

Predictor        Coef        StDev           T          P
Constant       1.8987       0.1830       10.38      0.000
Group          0.7168       0.2587        2.77      0.020

S = 0.4482     R-Sq = 43.4%    R-Sq(adj) = 37.8%

Analysis of Variance

Source            DF          SS          MS          F          P
Regression         1      1.5416      1.5416       7.68      0.020
Residual Error    10      2.0084      0.2008
Total             11      3.5500
```

The output indicates a significant group effect ($t = 2.77$, $p = 0.020$). The size of this group effect is estimated as the coefficient $\hat{\beta}_1 = 0.7168$. This means that the toluene group has an estimated mean value that is 0.7168 units higher than the mean value in the control group. The reader might wish to check that this calculation is correct, and that the t test given by the regression routine does actually give the same results as a t test performed according to textbook formulas. Also note that the F test in the output is related to the t test through $t^2 = F$: $2.77^2 = 7.68$. These two tests are identical.

4.8.4 One-way ANOVA

The generalization of t-test models to more than two groups is rather straightforward; we would need one new dummy variable for each new group. This leads to a simple one-way analysis of variance (ANOVA) model. Thus, a one-way ANOVA model with three treatments, each with two observations per treatment, can be written as

$$y_{ij} = \mu + \beta_j + e_{ij},$$
$$i = 1, 2, \ j = 1, 2, 3$$

We can introduce three dummy variables d_1, d_2 and d_3 such that

$$d_j = \begin{cases} 1 & \text{for group } j \\ 0 & \text{otherwise} \end{cases}.$$

4 General linear models

The model can now be written as

$$y_{ij} = \mu + \beta_1 d_{i1} + \beta_2 d_{i2} + \beta_3 d_{i3} + e_{ij}$$
$$= \mu + \sum \beta_j d_{ij} + e_{ij},$$
$$i = 1, 2, \; j = 1, 2, 3$$

Note that the third dummy variable d_3 is not needed. If we know the values of d_1 and d_2 the group membership is known, so d_3 is redundant and can be removed from the model. In fact, any combination of two of the dummy variables is sufficient for identifying group membership so the choice to delete one of them is to some extent arbitrary. After removing d_3, the model can be written as

$$y_{11} = \mu + \beta_1 \cdot 1 + \beta_2 \cdot 0 + e_{11}$$
$$y_{21} = \mu + \beta_1 \cdot 1 + \beta_2 \cdot 0 + e_{21}$$
$$y_{12} = \mu + \beta_1 \cdot 0 + \beta_2 \cdot 1 + e_{12}$$
$$y_{22} = \mu + \beta_1 \cdot 0 + \beta_2 \cdot 1 + e_{22}$$
$$y_{13} = \mu + \beta_1 \cdot 0 + \beta_2 \cdot 0 + e_{13}$$
$$y_{23} = \mu + \beta_1 \cdot 0 + \beta_2 \cdot 0 + e_{23}$$

Although there are three treatments we have only included two dummy variables for the treatments, i.e. we have chosen the restriction $\beta_3 = 0$.

Follow-up analyses

One of the results from a one-way ANOVA is an over-all F test of the hypothesis that all group (treatment) means are equal. If this test is significant, it can be followed up by various types of comparisons between the groups. Since the ANOVA provides an estimator $\hat{\sigma}_e^2 = MS_e$ of the residual variance σ_e^2, this estimator should be used in such group comparisons if the assumption of equal variance seems tenable.

A pairwise comparison between two group means, i.e. a test of the hypothesis that two groups have equal mean values, can be obtained as

$$t = \frac{\bar{y}_j - \bar{y}_{j'}}{\sqrt{MS_e \left(\frac{1}{n_j} + \frac{1}{n_{j'}}\right)}}$$

with degrees of freedom taken from MS_e. A confidence interval for the difference between the mean values can be obtained analogously.

In some cases it may be of interest to make comparisons which are not simple pairwise comparisons. For example, we may want to compare treatment 1 with the average of treatments 2, 3 and 4. We can then define a contrast in the treatment means as $L = \mu_1 - \frac{\mu_2+\mu_3+\mu_4}{3}$. A general way to write a contrast is

$$L = \sum_j h_j \mu_j,$$

where we define the weights h_j such that $\sum_j h_j = 0$. The contrast can be estimated as

$$\widehat{L} = \sum_j h_j \bar{y}_j,$$

and the estimated variance of $\widehat{L}$ is

$$\widehat{Var}\left(\widehat{L}\right) = MS_e \sum_j \frac{h_j^2}{n_j}.$$

This can be used for tests and confidence intervals on contrasts.

Example 4.4 Liss et al (1996) studied the effects of seven contrast media (used in X-ray investigations) on different physiological functions of 57 rats. One variable that was studied was the urine production. Table 4.4 shows the change in urine production of each rat before and after treatment with each medium. It is of interest to compare the contrast media with respect to the change in urine production.

This analysis is a one-way ANOVA situation. The procedure GLM in SAS produced the following result:

```
                        General Linear Models Procedure

Dependent Variable: DIFF    DIFF
                                  Sum of                Mean
Source               DF          Squares              Square    F Value    Pr > F
Model                 6       1787.9722541         297.9953757    16.46     0.0001
Error                50        905.1155428          18.1023109
Corrected Total      56       2693.0877969

              R-Square                  C.V.          Root MSE           DIFF Mean
              0.663912               61.95963         4.2546811           6.8668596

Source               DF        Type III SS        Mean Square    F Value    Pr > F
MEDIUM                6       1787.9722541         297.9953757    16.46     0.0001
```

There are clearly significant differences between the media ($p < 0.0001$). To find out more about the nature of these differences we requested Proc GLM to print estimates of the parameters, i.e. estimates of the coefficients β_i for each of the dummy variables. The following results

4 General linear models

Table 4.4: Change in urine production following treatment with different contrast media ($n = 57$).

Medium	Diff	Medium	Diff	Medium	Diff
Diatrizoate	32.92	Isovist	2.44	Ringer	0.10
Diatrizoate	25.85	Isovist	0.87	Ringer	0.40
Diatrizoate	20.75	Isovist	−0.22	Mannitol	9.19
Diatrizoate	20.38	Isovist	1.52	Mannitol	0.79
Diatrizoate	7.06	Omnipaque	8.51	Mannitol	10.22
Hexabrix	6.47	Omnipaque	16.11	Mannitol	4.78
Hexabrix	5.63	Omnipaque	7.22	Mannitol	14.64
Hexabrix	3.08	Omnipaque	9.03	Mannitol	6.98
Hexabrix	0.96	Omnipaque	10.11	Mannitol	7.51
Hexabrix	2.37	Omnipaque	6.77	Mannitol	9.55
Hexabrix	7.00	Omnipaque	1.16	Mannitol	5.53
Hexabrix	4.88	Omnipaque	16.11	Ultravist	12.94
Hexabrix	1.11	Omnipaque	3.99	Ultravist	7.30
Hexabrix	4.14	Omnipaque	4.90	Ultravist	15.35
Isovist	2.10	Ringer	0.07	Ultravist	6.58
Isovist	0.77	Ringer	−0.03	Ultravist	15.68
Isovist	−0.04	Ringer	0.34	Ultravist	3.48
Isovist	4.80	Ringer	0.08	Ultravist	5.75
Isovist	2.74	Ringer	0.51	Ultravist	12.18

were obtained:

```
                                         T for H0:    Pr > |T|    Std Error of
Parameter                 Estimate      Parameter=0                 Estimate

INTERCEPT               9.90787500 B        6.59       0.0001      1.50425691
MEDIUM    Diatrizoate  11.48412500 B        4.73       0.0001      2.42554139
          Hexabrix     -5.94731944 B       -2.88       0.0059      2.06740338
          Isovist      -8.24365278 B       -3.99       0.0002      2.06740338
          Mannitol     -2.21920833 B       -1.07       0.2882      2.06740338
          Omnipaque    -1.51817500 B       -0.75       0.4554      2.01817243
          Ringer       -9.69787500 B       -4.40       0.0001      2.20200665
          Ultravist     0.00000000 B          .           .             .

NOTE: The X'X matrix has been found to be singular and a generalized inverse
      was used to solve the normal equations.  Estimates followed by the
      letter 'B' are biased, and are not unique estimators of the parameters.
```

Note that Proc GLM reports the $\mathbf{X'X}$ matrix to be singular. This is as expected for an ANOVA model: not all dummy variables can be included in the model. (See Section 4.13 for a summary of the use of matrix algebra in general linear models.) The procedure excludes the last dummy variable, setting the parameter for Ultravist to 0. All

Table 4.5: Least squares means, and pairwise comparisons between treatments, for the contrast media experiment.

	Diatrizoate	Ultravist	Omnipaque	Mannitol	Hexabrix	Isovist	Ringer
Mean	21.39	9.91	8.39	7.69	3.96	1.66	0.21
Diatrizoate	—						
Ultravist	*	—					
Omnipaque	*	n.s.	—				
Mannitol	*	n.s.	n.s.	—			
Hexabrix	*	n.s.	n.s.	n.s.	—		
Isovist	*	*	*	n.s.	n.s.	—	
Ringer	*	*	*	*	n.s.	n.s.	—

other estimates are comparisons of the estimated mean value for that medium, with the mean value for Ultravist. Least squares estimates of the mean values for the media can be calculated and compared. Since this can result in a large number of pairwise comparisons (in this case, $7 \cdot 6/2 = 21$ comparisons), some method for protection against mass significance might be considered. The least squares means are given in Table 4.5 along with indications of significant pairwise differences using Bonferroni adjustment.

Before we close this example, we should take a look at how the data behave. For example, we can prepare a boxplot of the distributions for the different media. This boxplot is given in Figure 4.2. The plot indicates that the variation is quite different for the different media, with a large variation for diatrizoate and a small variation for ringer (which is actually a placebo). This suggests that one assumption underlying the analysis, the assumption of equal variance, may be violated. Another way of analyzing these data is presented on Page 114.

4.8.5 ANOVA: Factorial experiments

The ideas used above can be extended to factorial experiments that include more than one factor and possible interactions. The dummy variables that correspond to the interaction terms would then be constructed by multiplying the corresponding main effect dummy variables with each other.

This feature can be illustrated by considering a factorial experiment with factor A (two levels) and factor B (three levels), and where we have

4 General linear models

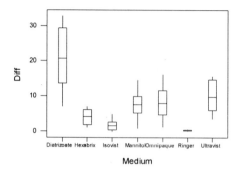

Figure 4.2: Boxplot of change in urine production for different contrast media.

two observations for each factor combination. The model is

$$y_{ijk} = \mu + \alpha_i + \beta_j + (\alpha\beta)_{ij} + e_{ijk},$$
$$i = 1, 2, \; j = 1, 2, 3, \; k = 1, 2$$

The number of dummy variables that we have included for each factor is equal to the number of factor levels minus one, i.e. the last dummy variable for each factor has been excluded. The number of non-redundant dummy variables equals the number of degrees of freedom for the effect. The model can be written as follows:

$$y_{111} = \mu + \alpha_1 \cdot 1 + \beta_1 \cdot 1 + \beta_2 \cdot 0 + (\alpha\beta)_{11} \cdot 1 + (\alpha\beta)_{12} \cdot 0 + e_{111}$$
$$y_{112} = \mu + \alpha_1 \cdot 1 + \beta_1 \cdot 1 + \beta_2 \cdot 0 + (\alpha\beta)_{11} \cdot 1 + (\alpha\beta)_{12} \cdot 0 + e_{112}$$
$$y_{121} = \mu + \alpha_1 \cdot 1 + \beta_1 \cdot 0 + \beta_2 \cdot 1 + (\alpha\beta)_{11} \cdot 0 + (\alpha\beta)_{12} \cdot 1 + e_{121}$$
$$y_{122} = \mu + \alpha_1 \cdot 1 + \beta_1 \cdot 0 + \beta_2 \cdot 1 + (\alpha\beta)_{11} \cdot 0 + (\alpha\beta)_{12} \cdot 1 + e_{122}$$
$$y_{131} = \mu + \alpha_1 \cdot 1 + \beta_1 \cdot 0 + \beta_2 \cdot 0 + (\alpha\beta)_{11} \cdot 0 + (\alpha\beta)_{12} \cdot 0 + e_{131}$$
$$y_{132} = \mu + \alpha_1 \cdot 1 + \beta_1 \cdot 0 + \beta_2 \cdot 0 + (\alpha\beta)_{11} \cdot 0 + (\alpha\beta)_{12} \cdot 0 + e_{132}$$
$$y_{211} = \mu + \alpha_1 \cdot 0 + \beta_1 \cdot 1 + \beta_2 \cdot 0 + (\alpha\beta)_{11} \cdot 1 + (\alpha\beta)_{12} \cdot 0 + e_{211}$$
$$y_{212} = \mu + \alpha_1 \cdot 0 + \beta_1 \cdot 1 + \beta_2 \cdot 0 + (\alpha\beta)_{11} \cdot 1 + (\alpha\beta)_{12} \cdot 0 + e_{212}$$
$$y_{221} = \mu + \alpha_1 \cdot 0 + \beta_1 \cdot 0 + \beta_2 \cdot 1 + (\alpha\beta)_{11} \cdot 0 + (\alpha\beta)_{12} \cdot 1 + e_{221}$$
$$y_{222} = \mu + \alpha_1 \cdot 0 + \beta_1 \cdot 0 + \beta_2 \cdot 1 + (\alpha\beta)_{11} \cdot 0 + (\alpha\beta)_{12} \cdot 1 + e_{222}$$
$$y_{231} = \mu + \alpha_1 \cdot 0 + \beta_1 \cdot 0 + \beta_2 \cdot 0 + (\alpha\beta)_{11} \cdot 0 + (\alpha\beta)_{12} \cdot 0 + e_{231}$$
$$y_{232} = \mu + \alpha_1 \cdot 0 + \beta_1 \cdot 0 + \beta_2 \cdot 0 + (\alpha\beta)_{11} \cdot 0 + (\alpha\beta)_{12} \cdot 0 + e_{232}$$

Example 4.5 *Lindahl et al (1999) studied certain reactions of fungal*

Table 4.6: Data for a two-factor experiment.

Species	Size	C	Species	Size	C
H	Large	0.0010	S	Large	0.0021
H	Large	0.0011	S	Large	0.0001
H	Large	0.0017	S	Large	0.0016
H	Large	0.0008	S	Large	0.0046
H	Large	0.0010	S	Large	0.0035
H	Large	0.0028	S	Large	0.0065
H	Large	0.0003	S	Large	0.0073
H	Large	0.0013	S	Large	0.0039
H	Small	0.0061	S	Small	0.0007
H	Small	0.0010	S	Small	0.0011
H	Small	0.0020	S	Small	0.0019
H	Small	0.0018	S	Small	0.0022
H	Small	0.0033	S	Small	0.0011
H	Small	0.0015	S	Small	0.0012
H	Small	0.0040	S	Small	0.0009
H	Small	0.0041	S	Small	0.0040

mycelia on pieces of wood by using radioactively labeled ^{32}P. In one of the experiments, two species of fungus (Paxillus involutus and Suillus variegatus) were used, along with two sizes of wood pieces (large and small); the response was a certain chemical measurement denoted by C. The data are reproduced in Table 4.6.

These data were analyzed as a factorial experiment with two factors. Part of the Minitab output was:

General Linear Model: C versus Species; Size

```
Analysis of Variance for C, using Adjusted SS for Tests

Source         DF      Seq SS      Adj SS      Adj MS       F      P
Species         1   0.0000025   0.0000025   0.0000025    0.93  0.342
Size            1   0.0000002   0.0000002   0.0000002    0.09  0.772
Species*Size    1   0.0000287   0.0000287   0.0000287   10.82  0.003
Error          28   0.0000742   0.0000742   0.0000027
Total          31   0.0001056
```

The main conclusion from this analysis is that the interaction species × size is highly significant. This means that the effect of size is different for different species. In such cases, interpretation of the main effects is not very meaningful. As a tool for interpreting the interaction effect, a so called interaction plot can be prepared. Such a plot for these data is as given in Figure 4.3. The mean value of the response for species S is higher for large pieces of wood than for small oness. For species H the

4 General linear models

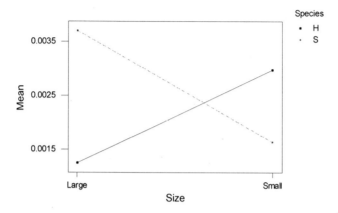

Figure 4.3: Interaction plot for the 2-factor experiment.

opposite is true: the mean value is larger for small pieces of wood. This is an example of interaction.

4.8.6 Analysis of covariance

In regression analysis models the data contain quantitative independent variables. In ANOVA models, the data contain dummy variables corresponding to treatments, design structure and possible interactions. It is quite possible to include a mixture of quantitative variables and dummy variables in the data. Such models are called covariance analysis, or ANCOVA, models.

Let us look at a simple case where there are two groups and one covariate. Several different models can be considered for the analysis of such data even in the simple case where we assume that all relationships are linear:

1. There is no relationship between x and y in any of the groups and the groups have the same mean value.

2. There is a relationship between x and y; the relationship is the same in the groups.

3. There is no relationship between x and y but the groups have different levels.

Some applications

4. There is a relationship between x and y; the lines are parallel but at different levels.

5. There is a relationship between x and y; the lines are different in the groups.

These five cases correspond to the different models that are presented below:

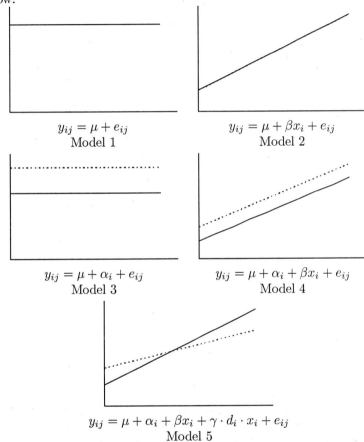

$$y_{ij} = \mu + e_{ij}$$
Model 1

$$y_{ij} = \mu + \beta x_i + e_{ij}$$
Model 2

$$y_{ij} = \mu + \alpha_i + e_{ij}$$
Model 3

$$y_{ij} = \mu + \alpha_i + \beta x_i + e_{ij}$$
Model 4

$$y_{ij} = \mu + \alpha_i + \beta x_i + \gamma \cdot d_i \cdot x_i + e_{ij}$$
Model 5

Model 5 is the most general of the models, allowing for different intercepts ($\mu + \alpha_i$) and different slopes $\beta + \gamma d_i$, where d is a dummy variable indicating group membership. If it can be assumed that the term γd_i is zero for all i, then we are back at model 4. If, in addition, all α_i are zero, then model 2 is correct. If, on the other hand, β is zero, we would use model 3. If finally β is zero in model 2, then model 1 describes the situation. This is an example of a set of models where some of the models are nested within other models. The model choice can be made

4 General linear models

by comparing any model to a simpler model which only differs in terms of one factor.

4.9 Estimability

In some types of general linear models it is impossible to estimate all model parameters. It is then necessary to restrict some parameters to being zero, or to use some other restriction on the parameters.

As an example, a two-factor ANOVA model with two levels of factor A, three levels of factor B and two replications can be written as

$$y_{ijk} = \mu + \alpha_i + \beta_j + (\alpha\beta)_{ij} + e_{ijk},$$
$$i = 1, 2, \; j = 1, 2, 3, \; k = 1, 2.$$

In this model it would be possible to replace μ with $\mu + c$ and to replace each α_i with $\alpha_i - c$, where c is some constant. The same kind of ambiguity holds also for other parameters of the model. This model contains a total of 12 parameters: μ, α_1, α_2, β_1, β_2, β_3, $(\alpha\beta)_{11}$, $(\alpha\beta)_{12}$, $(\alpha\beta)_{13}$, $(\alpha\beta)_{21}$, $(\alpha\beta)_{22}$, and $(\alpha\beta)_{23}$, but only 6 of the parameters can be estimated. As noted above, computer programs often solve this problem by restricting some parameters to being zero.

However, it may be possible to estimate certain functions of the parameters in a unique way. Such functions, if they exist, are called estimable functions. A linear combination of model parameters is estimable if it can be written as a linear combination of expected values of the observations.

Let us denote with $\mu_{ij.}$ the mean value for the treatment combination that has factor A at level i and factor B at level j. It holds that

$$\mu_{ij.} = E\left(y_{ijk}\right) = \mu + \alpha_i + \beta_j + (\alpha\beta)_{ij}$$

which is a linear function of the parameters. This function is estimable. In addition, any linear function of the $\mu_{ij.}$:s is also estimable. For example, the expected value of all observations with factor A at level i can be written as

$$\mu_{i..} = \frac{\mu_{i1.} + \mu_{i2.} + \mu_{i3.}}{3}.$$

This is a linear function of cell means. Since the cell means are estimable, $\mu_{i..}$ is also estimable.

4.10 Assumptions in general linear models

The classical application of general linear models rests on the following set of assumptions:

1. The model used for the analysis is assumed to be correct.

2. The residuals are assumed to be independent.

3. The residuals are assumed to follow a normal distribution.

4. The residuals are assumed to have the same variance σ_e^2, independent of **x**, i.e. the residuals are homoscedastic.

Different diagnostic tools have been developed to detect departures from these assumptions; these are discussed on page 25.

4.11 Model building

4.11.1 Computer software for GLMs

There are many options for fitting general linear models to data. One option is to use a regression package and leave it to the user to construct appropriate dummy variables for class variables. However, most statistical packages have routines for general linear models that automatically construct the appropriate set of dummy variables.

Let us use letters at the end of the alphabet (x, y, z) to denote numeric variables. y will be used for the dependent variable. Letters at the beginning of the alphabet (A, B) will symbolize class variables (groups, treatments, blocks, etc.).

Computer software requires the user to state the model in symbolic terms. The model statement contains operators that specify different aspects of the model. In the following table we list the operators used by SAS. Examples of the use of the operators are given below.

Operator	Explanation, SAS example
*	Interaction: A*B. Also used for polynomials: X*X
(none)	Both effects present: A B
|	All main effects and interactions: A|B=A B A*B
()	Nested factor: A(B). "A nested within B"
@	Order operator: A|B|C @ 2 means that all main effects and all interaction up to and including second order interactions are included.

4 General linear models

The kinds of models that we have discussed in this chapter can symbolically be written in SAS language as indicated in the following table.

Model	Computer model (SAS)
Simple linear regression	Y = X
Multiple regression	Y = X Z
t tests, one-way ANOVA	Y = A
Two-way ANOVA with interaction	Y = A B A*B or Y=A|B
Covariance analysis model 1	Y =
Covariance analysis model 2	Y = X
Covariance analysis model 3	Y = A
Covariance analysis model 4	Y = A X
Covariance analysis model 5	Y = A X A*X

4.11.2 Model building strategy

Statistical model building is an art as much as it is a science. There are many requirements on models: they should make sense from a subject-matter point of view, they should be simple, and at the same time they should capture most of the information in the data. A good model is a compromise between parsimony and completeness. This means that it is impossible to state simple rules for model building: there will certainly be cases where the rules are not relevant. However, the following suggestions, based on McCullagh and Nelder (1989, p. 89), are useful in many cases:

- Include all relevant main effects in the model, even those that are not significant.

- If an interaction is included, the model should also include all main effects and interactions it comprises. For example, if the interaction A*B*C is included, the model should also include A, B, C, A*B, A*C and B*C.

- A model that contains polynomial terms of type x^a should also contain the lower-degree terms x, x^2, ... , x^{a-1}.

- Covariates that do not have any detectable effect should be excluded.

- The conventional 5% significance level is often too strict for model building purposes. A significance level in the range 15-25% may be used instead.

- Alternatively, criteria like the Akaike information criterion (AIC) can be used. The AIC is described on page 202.

4.12 A covariance analysis example

We will consider a small example where "time" serves the role of a covariate.

Example 4.6 *Below are some data on the emission of carbon dioxide from the root systems of plants (Zagal et al, 1993). Two levels of nitrogen were used, and samples of plants were analyzed 24, 30, 35 and 38 days after germination. The data were as follows:*

	Days from germination			
Level of nitrogen	24	30	35	38
High	8.220	19.296	25.479	31.186
	12.594	31.115	34.951	39.237
	11.301	18.891	20.688	21.403
Low	15.255	28.200	32.862	41.677
	11.069	26.765	34.730	43.448
	10.481	28.414	35.830	45.351

4.12.1 Two-way ANOVA without interaction

The structure of the example data is a two-factor analysis of variance design, with level and days as factors. We can use our favourite statistical package for the analysis; in these notes we will mainly use SAS and Minitab. We start by looking at an analysis that does not contain any interaction term. If we regard "days" as a classification factor, we get the following result:

General Linear Model: CO2 versus Nitrogen; Days

```
Analysis of Variance for CO2, using Adjusted SS for Tests

Source     DF    Seq SS    Adj SS    Adj MS       F      P
Nitrogen    1    264.81    264.81    264.81   10.03  0.005
Days        3   2133.25   2133.25    711.08   26.93  0.000
Error      19    501.60    501.60     26.40
Total      23   2899.66
```

The results indicate a significant effect of days, and a significant effect of level of nitrogen. In this analysis we have not used the fact

4 General linear models

that "days" is a numeric variable. A graph of the relation between the response and days is given in Figure 4.4.

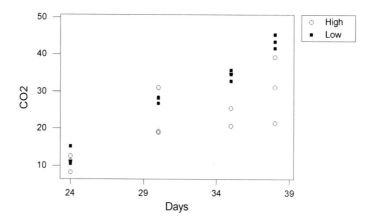

Figure 4.4: Relation between CO_2 emission and time.

The graph indicates a fairly steady increase in CO2 level over time, for both treatments. This suggests that it may be reasonable to include "days" in the model as a quantitative factor. A model that includes a linear term in time gives the following result:

General Linear Model: CO2 versus Nitrogen

```
Analysis of Variance for CO2, using Adjusted SS for Tests

Source      DF    Seq SS    Adj SS    Adj MS       F        P
Nitrogen     1     264.8     264.8     264.8   10.13    0.004
Days         1    2085.6    2085.6    2085.6   79.74    0.000
Error       21     549.2     549.2      26.2
Total       23    2899.7
```

It would also be possible to include quadratic and cubic terms in the model. In general, if there are t time points, the model can contain terms in time up to the $(t-1)$:th degree. The analyses are summarised in Table 4.7.

We note that the model with a cubic term in D gives results that are identical to the analysis where D is regarded as qualitative. Also, the terms D, D^2 and D^3 each have one $d.f.$ adding up to 3 $d.f.$ which is

A covariance analysis example

Table 4.7: Summary of different analyses of the CO2 data.

Source	D qualitative	Linear	Quadratic	Cubic
Nitrogen	264.81	264.81	264,81	264.81
D	2133.25	2085.61	2085.61	2085.61
D^2			19.10	19.10
D^3				28.53
Sum days	*2133.25*	*2085.61*	*2104.72*	*2133.25*
Error	501.60	549.20	530.13	501.60
Total	2899.60	2899.60	2899.60	2899.60

the same as the $d.f$ for D in the qualitative model. This indicates that a model with t measurements in time can be analyzed with time as a qualitative factor with $(t-1)$ d.f., or as a polynomial in time up to the $(t-1)$:th degree. The $(t-1)$ d.f. of the time factor can be split up into linear, quadratic, cubic terms, up to terms of degree $(t-1)$.

It may also be possible to simplify the model by including only terms of a lower degree than $(t-1)$. In the cubic model, the cubic term is not significant ($F = 1.08$; $p = 0.312$) so it does not contribute significantly to the model. This test was calculated as

$$F = \frac{SS_3}{MS_e} = \frac{28.53}{501.60/19} = 1.08.$$

In the quadratic model, the second degree term is not significant ($F = 0.72$; $p = 0.406$). This suggests that it may be enough to include the linear term in the model. The hypothesis testing strategy is to test "backwards": start with the term with power $(t-1)$. If this is not significant, simplify the model by removing this term and test the term with power $(t-2)$, and so on.

Before we settle on a final model we should investigate whether there is any interaction between nitrogen and days.

4.12.2 A model with interaction

A model with interaction and with days as a qualitative factor gives the following result:

General Linear Model: CO2 versus Nitrogen; Days

```
Analysis of Variance for CO2, using Adjusted SS for Tests

Source          DF    Seq SS    Adj SS    Adj MS      F       P
Nitrogen         1    264.81    264.81    264.81   10.65   0.005
Days             3   2133.25   2133.25    711.08   28.60   0.000
Nitrogen*Days    3    103.76    103.76     34.59    1.39   0.282
Error           16    397.84    397.84     24.86
Total           23   2899.66
```

These results do not indicate any significant interaction between nitrogen and days, i.e. the process seems to proceed at the same rate for both levels of nitrogen. However, the graph (page 94) shows a tendency that the slope may be somewhat larger for the low nitrogen level. This would be detected as an interaction between the qualitative factor nitrogen and the quantitative factor days. Such an analysis is:

General Linear Model: CO2 versus Nitrogen

```
Analysis of Variance for CO2, using Adjusted SS for Tests

Source          DF    Seq SS    Adj SS    Adj MS      F       P
Nitrogen         1    264.81     47.79     47.79    2.10   0.163
Days             1   2085.61   2085.61   2085.61   91.76   0.000
Nitrogen*Days    1     94.67     94.67     94.67    4.17   0.055
Error           20    454.57    454.57     22.73
Total           23   2899.66
```

The interaction is not quite significant at the 5% level. If the main purpose of the analysis is to find a predictive model, we would choose to retain the interaction in the model.

It is quite possible to include quadratic and cubic terms in the model. These terms may also interact with nitrogen. A number of possible analyses are summarized in Table 4.8.

There are no indications of significant interaction with the second or third degree terms. Also, these terms are not significant as such.

A simple way of choosing among the many models we have discussed is to compare the value of the Akaike Information Criterion (AIC) for the different models. Table 4.9 summarizes the AIC values. The model with "days" as a numeric variable and that includes an interation term has the smallest AIC value, 137.8.

Thus, a model that provides a reasonable approximation to these data would include the effect of nitrogen, the linear effect of time, and

Table 4.8: Models with and without interaction for the CO_2 data.

Source	Qualitative +interaction	Linear +interaction	Quadratic +interaction	Cubic +interaction
Nitrogen	264.81	264.81	264.81	264.81
D	2133.25	2085.61	2085.61	2085.61
D2			19.10	19.10
D3				28.53
Sum D	2133.25	2085.61	2104.71	2133.25
N*D	103.76	94.67	94.67	94.67
N*D2			6.19	6.19
N*D3				2.91
Sum	103.76	94.67	100.86	103.76
Error	397.84	454.57	429.27	397.84
Total	2899.66	2899.66	2899.66	2899.66

Table 4.9: AIC values for different models

	No interaction	With interaction
"Class"	150.3	154.1
Linear	141.6	137.8
Quadratic	145.1	143.9
Cubic	150.3	154.1

the interaction between N and time. Figure 4.5 illustrates this model, along with the raw data.

4.13 General linear models using matrix algebra

General linear models may be written, in a compact way, using matrix algebra. Some programs for GLM's use matrix algebra terminology in their output; see for example page 84. It may therefore be useful to describe the use of matrix algebra for GLM's. A brief introduction to matrix algebra is given in the Appendix (page 287).

A general linear model with p x-variables and n observations may be written as

$$y_i = \beta_0 + \beta_1 x_{i1} + \ldots + \beta_p x_{ip} + e_i$$

In matrix terms,

4 General linear models

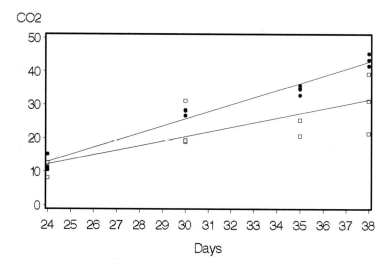

Figure 4.5: A model for the CO_2 data: linear trends with different slope for the two groups.

$$\mathbf{y} = \mathbf{X}\boldsymbol{\beta} + \mathbf{e}, \text{ i.e.}$$

$$\begin{pmatrix} y_1 \\ y_2 \\ \vdots \\ y_n \end{pmatrix} = \begin{pmatrix} 1 & x_{11} & \cdots & x_{1p} \\ 1 & x_{21} & & \\ \vdots & & \ddots & \\ 1 & x_{n1} & & x_{np} \end{pmatrix} \begin{pmatrix} \beta_0 \\ \beta_1 \\ \vdots \\ \beta_p \end{pmatrix} + \begin{pmatrix} e_1 \\ e_2 \\ \vdots \\ e_n \end{pmatrix}$$

Here, $\mathbf{y}$ is an $n \times 1$ vector of observations on the dependent variable, $\mathbf{X}$ is a known design matrix of dimension $n \times (p+1)$ that contains data on the x-variables, and one column of 1:s, corresponding to the intercept, $\boldsymbol{\beta}$ is a vector of size $(p+1) \times 1$ that contains parameters to be estimated, and $\mathbf{e}$ is an $n \times 1$ vector of residuals.

Estimation of the parameters in $\boldsymbol{\beta}$ is made, using the method of least squares, by minimizing the residual sum of squares

$$\mathbf{e}'\mathbf{e} = (\mathbf{y} - \mathbf{X}\boldsymbol{\beta})'(\mathbf{y} - \mathbf{X}\boldsymbol{\beta}).$$

with respect to the elements in $\boldsymbol{\beta}$. This gives the so-called normal

General linear models using matrix algebra

equations as

$$\mathbf{X}'\mathbf{X}\boldsymbol{\beta} = \mathbf{X}'\mathbf{y}.$$

If the matrix $\mathbf{X}'\mathbf{X}$ is nonsingular, this yields as estimators of the parameters of the model

$$\widehat{\boldsymbol{\beta}} = (\mathbf{X}'\mathbf{X})^{-1}\mathbf{X}'\mathbf{y}.$$

If the inverse of $\mathbf{X}'\mathbf{X}$ does not exist, we can still find a solution although it may not be unique. We can use generalized inverses and find a solution as

$$\widehat{\boldsymbol{\beta}} = (\mathbf{X}'\mathbf{X})^{-}\mathbf{X}'\mathbf{y}.$$

alternatively we can restrict the number of parameters in the model by introducing constraints that lead to a nonsingular $\mathbf{X}'\mathbf{X}$.

Some examples

Linear regression A simple linear regression with $n = 4$ observations and and $p = 1$ variables is written as

$$\begin{pmatrix} y_1 \\ y_2 \\ y_3 \\ y_4 \end{pmatrix} = \begin{pmatrix} 1 & x_1 \\ 1 & x_2 \\ 1 & x_3 \\ 1 & x_4 \end{pmatrix} \begin{pmatrix} \beta_0 \\ \beta_1 \end{pmatrix} + \begin{pmatrix} e_1 \\ e_2 \\ e_3 \\ e_4 \end{pmatrix}.$$

Multiple regression A multiple linear regression with $n = 6$ observations and $p = 2$ x-variables is

$$\begin{pmatrix} y_1 \\ y_2 \\ y_3 \\ y_4 \\ y_5 \\ y_6 \end{pmatrix} = \begin{pmatrix} 1 & x_{11} & x_{12} \\ 1 & x_{21} & x_{22} \\ 1 & x_{31} & x_{32} \\ 1 & x_{41} & x_{42} \\ 1 & x_{51} & x_{52} \\ 1 & x_{61} & x_{62} \end{pmatrix} \begin{pmatrix} \beta_0 \\ \beta_1 \\ \beta_2 \end{pmatrix} + \begin{pmatrix} e_1 \\ e_2 \\ e_3 \\ e_4 \\ e_5 \\ e_6 \end{pmatrix}.$$

t test A two-sample t test with three observations per treatment can be written

4 General linear models

$$\begin{pmatrix} y_{11} \\ y_{12} \\ y_{13} \\ y_{21} \\ y_{22} \\ y_{23} \end{pmatrix} = \begin{pmatrix} 1 & 1 \\ 1 & 1 \\ 1 & 1 \\ 1 & 0 \\ 1 & 0 \\ 1 & 0 \end{pmatrix} \begin{pmatrix} \mu \\ \beta \end{pmatrix} + \begin{pmatrix} e_{11} \\ e_{12} \\ e_{13} \\ e_{21} \\ e_{22} \\ e_{23} \end{pmatrix}.$$

The second column in **X** contains values of the single dummy variable.

One-way ANOVA A one-way ANOVA model with three treatments and two observations per treatment is

$$\begin{pmatrix} y_{11} \\ y_{21} \\ y_{12} \\ y_{22} \\ y_{13} \\ y_{23} \end{pmatrix} = \begin{pmatrix} 1 & 1 & 0 \\ 1 & 1 & 0 \\ 1 & 0 & 1 \\ 1 & 0 & 1 \\ 1 & 0 & 0 \\ 1 & 0 & 0 \end{pmatrix} \begin{pmatrix} \mu \\ \beta_1 \\ \beta_2 \end{pmatrix} + \begin{pmatrix} e_{11} \\ e_{21} \\ e_{12} \\ e_{22} \\ e_{13} \\ e_{23} \end{pmatrix}$$

The first column in **X** corresponds to the intercept term $\mu = \beta_0$, the remaining two columns to the two dummy variables.

Two-factor ANOVA A two-factor ANOVA model with two levels of factor A, three levels of factor B and two observations per treatment combination is

$$\begin{pmatrix} y_{111} \\ y_{112} \\ y_{121} \\ y_{122} \\ y_{131} \\ y_{132} \\ y_{211} \\ y_{212} \\ y_{221} \\ y_{222} \\ y_{231} \\ y_{232} \end{pmatrix} = \begin{pmatrix} 1 & 1 & 1 & 0 & 1 & 0 \\ 1 & 1 & 1 & 0 & 1 & 0 \\ 1 & 1 & 0 & 1 & 0 & 1 \\ 1 & 1 & 0 & 1 & 0 & 1 \\ 1 & 1 & 0 & 0 & 0 & 0 \\ 1 & 1 & 0 & 0 & 0 & 0 \\ 1 & 0 & 1 & 0 & 0 & 0 \\ 1 & 0 & 1 & 0 & 0 & 0 \\ 1 & 0 & 0 & 1 & 0 & 0 \\ 1 & 0 & 0 & 1 & 0 & 0 \\ 1 & 0 & 0 & 0 & 0 & 0 \\ 1 & 0 & 0 & 0 & 0 & 0 \end{pmatrix} \begin{pmatrix} \mu \\ \alpha_1 \\ \beta_1 \\ \beta_2 \\ (\alpha\beta)_{11} \\ (\alpha\beta)_{12} \end{pmatrix} + \begin{pmatrix} e_{111} \\ e_{112} \\ e_{121} \\ e_{122} \\ e_{131} \\ e_{132} \\ e_{211} \\ e_{212} \\ e_{221} \\ e_{222} \\ e_{231} \\ e_{232} \end{pmatrix}.$$

4.14 GLM: Summary

General linear models (GLM) are used to study the relation between a response variable y and one or more independent variables (covariates) $x_1, x_2, \ldots x_p$. The covariates may be:

1. Numeric variables for which we assume a linear relation with y.

2. Categorical ("class") variables such as treatment code, block code, sex, etc).

Models with only class variables are often called Anova models (or t test models if there is only one binary variable). Variables with only numeric variables are (multiple) regression models. Models that contain both types of variables are called analysis of covariance models.

Programs for GLM will automatically translate variables coded as "class" variables, and interactions between these, into dummy variables.

SAS Proc GLM (and other procedures) in SAS will assume that all variables are numeric unless they are stated as "class" variables. Thus, the program
PROC GLM;
CLASS A;
MODEL y=A X;
RUN;
defines variable A as a class variable, while x is assumed to be numeric (since it is not mentioned in the CLASS list).

Minitab A GLM is defined in Minitab via the sequence Stat, Anova, General linear model. Class variables (and their interactions) are given in the `Model` box, numeric variables are given using the `Covariates` push button.

4 General linear models

4.15 Exercises

Exercise 4.1 *In an experiment on maize, six varieties (A, B, C, D, E and F) (Var) were tested in four blocks (Block). Unfortunately, the experiment was carried out in such a way that the plant density became rather uneven. It was therefore decided to include the plant density (stand) in the analysis. The following SAS program was used to analyse the data:*

```
DATA Sned18;
INPUT Var $ Stand Yield Block;
IF Var='A' THEN D1=1; ELSE D1=0;
IF Var='B' THEN D2=1; ELSE D2=0;
IF Var='C' THEN D3=1; ELSE D3=0;
IF Var='D' THEN D4=1; ELSE D4=0;
IF Var='E' THEN D5=1; ELSE D5=0;

IF Block=1 THEN B1=1; ELSE B1=0;
IF Block=2 THEN B2=1; ELSE B2=0;
IF Block=3 THEN B3=1; ELSE B3=0;

SD1=Stand*D1;
SD2=Stand*D2;
SD3=Stand*D3;
SD4=Stand*D4;
SD5=Stand*D5;
CARDS;
A   28  202  1
A   22  165  2

(Most of the data not shown here; total n=24)

F   24  204  4
;

PROC REG DATA=Sned18;
    MODEL Yield = Stand;                                            /*Model 1*/
    MODEL Yield = Stand D1 D2 D3 D4 D5;                             /*Model 2*/
    MODEL Yield = Stand D1 D2 D3 D4 D5 B1 B2 B3;                    /*Model 3*/
    MODEL Yield = Stand D1 D2 D3 D4 D5 B1 B2 B3 SD1 SD2 SD3 SD4 SD5; /*Model 4*/
RUN;
```

A. For each of the following statements, state which of the models (1, 2, 3 or 4) agrees with the statement:

(a) "The different varieties display different degrees of sensitivity to plant density."

(b) "The yield is the same for all varieties."

Selected parts of the outputs for the four models are given below. Using the outputs, answer the following questions:

B. Test whether the regression lines of Yield on Stand can be assumed to be parallel for the different varieties.

C. Assume that the lines are parallel. Construct an ANOVA table, where the total variation is subdivided into variety, stand, block and error.

D. Use your ANOVA table from C. to test the hypothesis that the different varieties give, on the average, the same yield.

Model: MODEL1
Dependent Variable: YIELD

Analysis of Variance

Source	DF	Sum of Squares	Mean Square	F Value	Prob>F
Model	1	12161.16728	12161.16728	41.051	0.0001
Error	22	6517.33272	296.24240		
C Total	23	18678.50000			

Model: MODEL2
Dependent Variable: YIELD

Analysis of Variance

Source	DF	Sum of Squares	Mean Square	F Value	Prob>F
Model	6	15814.81965	2635.80327	15.647	0.0001
Error	17	2863.68035	168.45179		
C Total	23	18678.50000			

Model: MODEL3
Dependent Variable: YIELD

Analysis of Variance

Source	DF	Sum of Squares	Mean Square	F Value	Prob>F
Model	9	17317.21407	1924.13490	19.789	0.0001
Error	14	1361.28593	97.23471		
C Total	23	18678.50000			

Model: MODEL4
Dependent Variable: YIELD

Analysis of Variance

Source	DF	Sum of Squares	Mean Square	F Value	Prob>F
Model	14	17729.75564	1266.41112	12.013	0.0004
Error	9	948.74436	105.41604		
C Total	23	18678.50000			

Exercise 4.2 *The relation between $y=$Yield and $x=$Amount of fertilizer has been studied for three geographical areas. The total number of observations is $n=60$. The following SAS program has been used for the analysis:*

4 General linear models

```
DATA Wheat;
INPUT Area X Y;
IF Area = 1 THEN D1=1; ELSE D1=0;
IF Area = 2 THEN D2=1; ELSE D2=0;
DX1 = D1 * X;
DX2 = D2 * X;
CARDS;

[The data are placed here]

;
PROC REG;
   MODEL Y = X D1 D2 DX1 DX2;          /*Model 1*/;
   MODEL Y = X D1 D2;                  /*Model 2*/;
   MODEL Y = X;                        /*Model 3*/;
RUN;
```

Selected parts of the output were:

Model 1:

$$Y = 2.0 + 4.8 \cdot D1 + 4.5 \cdot D2 + 1.0 \cdot X - 0.2 \cdot DX1 - 0.5 \cdot DX2$$

Anova table:

Source	DF	SS	MS
Regression		858	
Residual		142	
Total	59	1000	

Model 2:

$$Y = 3.5 + 3.5 \cdot D1 + 1.1 \cdot D2 + 0.7 \cdot X$$

Anova table:

Source	DF	SS	MS
Regression		847	
Residual		153	
Total	59	1000	

Model 3:

$$Y = 5.0 + 0.7 \cdot X$$

Anova table:

Source	DF	SS	MS
Regression		740	
Residual		260	
Total	59	1000	

A. Test if the regression lines for the three areas can be assumed to be parallel.

B. Test if the intercepts for the regression lines can be assumed to be equal for the three areas.

C. Which of the models (1, 2 or 3) do you consider to be the best descriptions of the relationship?

D. Draw a graph of the regression lines for the model you consider best. Use a ruler! The smallest value of X is 3, and the largest is 10. The regressions should be drawn using the parameter estimates reported above, and to scale.

Exercise 4.3 *The following data are from an experiment to study the relationship between the weight of turkeys, and various factors. The following data have been collected for 13 turkeys:*

X = *age in weeks*, Y = *weight in pounds, Origin* = *State where the turkey was raised:*

G = *Georgia*
V = *Virginia*
W = *Wisconsin*
D1 Dummy variable. D1=1 if Origin=G; D1=0 else
D2 Dummy variable. D2=1 if Origin=V; D2=0 else
The data are as follows:

Age X	Weight Y	Origin	D1	D2
28	13.3	G	1	0
20	8.9	G	1	0
32	15.1	G	1	0
22	10.4	G	1	0
29	13.1	V	0	1
27	12.4	V	0	1
28	13.2	V	0	1
26	11.8	V	0	1
21	11.5	W	0	0
27	14.2	W	0	0
29	15.4	W	0	0
23	13.1	W	0	0
25	13.8	W	0	0

A printout from a multiple regression analysis of Y on X, D1 and D2 is as follows:

4 General linear models

Regression Analysis

```
The regression equation is
Weight = 1.43 + 0.487 Age - 1.92 D1 - 2.19 D2

Predictor      Coef      Stdev    t-ratio       p
Constant     1.4309     0.6574       2.18   0.058
Age         0.48676    0.02574      18.91   0.000
D1          -1.9184     0.2018      -9.51   0.000
D2          -2.1919     0.2114     -10.37   0.000

s = 0.3002    R-sq = 97.9%    R-sq(adj) = 97.3%

Analysis of Variance

SOURCE       DF         SS         MS        F        p
Regression    3     38.606     12.869   142.78    0.000
Error         9      0.811      0.090
Total        12     39.417

SOURCE       DF     SEQ SS
Age           1     26.202
D1            1      2.717
D2            1      9.687
```

Answer the following questions using the output:

A. How much do turkeys grow per week, on the average?

B. Are there any significant differences in weight between turkeys from Wisconsin and turkeys from Georgia?

C. Give a point estimate of the average weight difference between turkeys form Georgia and turkeys from Virginia.

D. What weight would we expect for a turkey from Virginia of age 24 weeks?

E. Plot the data, together with the regression line(s) you can extract from the computer printout.

Exercise 4.4 In a 1974 study of 1072 men, a multiple regression was calculated to show how lung function was related to several factors, including some hazardous occupations. The estimated regression function was as follows:

$$Airc = 4500 \quad -39 \times Age \quad -9.0 \times Smok \quad -350 \times Chemw$$
$$(s.e.) \qquad\qquad (1.8) \qquad\quad (2.2) \qquad\qquad (46)$$
$$-380 \times Farmw \quad -180 \times Firew$$
$$(53) \qquad\qquad (54)$$

Explanations: The numbers given are the estimated regression coefficients for the different variables. The standard error (s.e.) of each coefficient is given in parenthesis.

Airc   Air capacity: Air (ml) expired in one second.
Age   Age in years
Smok   Number of cigarettes smoked per day
Chemw   1 if subject is a chemical worker, 0 if not
Farmw   1 if subject is a farm worker, 0 if not
Firew   1 if subject is a firefighter, 0 if not

A fourth occupation, physician, served as reference group and did not need a dummy.

Questions:

A. Test the null hypothesis that firefighters and physicians have equal average air capacity.

B. Estimate the difference in air capacity between chemical workers and farm workers. Which of the groups has the highest mean value?

C. Estimate the average air capacity of 50-year-old firefighters who smoke 15 cigarettes per day.

Exercise 4.5 Data in the separately supplied file come from a clinical experiment. Since the results have not yet been published, some details about treatments etc. cannot be revealed. However, the following facts are known:

The experiment concerns treatment in connection with a certain operation. During the operation, the patient can either get the traditional treatment (B) or a new treatment that may give fewer side effects (A). Treatment A might be less effective. However, in this analysis, we will not look at the actual effect of the treatment: we will only look at the side effects.

The patients were randomized on the two treatments A and B. For each patient, a certain laboratory test was taken before the operation (y_0) and 1 day after the operation (y_1). The variable y indicates the incidence of side effects of the treatment.

Treatment B means that a certain substance X is given to the patient. During the operation it turned out that some patients who were randomized to treatment A also needed some amount of X. The dose, however, was generally lower for these patients than for those that were randomized to X treatment (i.e. to treatment B). The dose of X for all patients (both A and B) is given in the variable x.

The data consist of the following variables:
Patient: Sequence number of the patient
Age: Age of the patient
Treat: Treatment (A or B)
x: Dose of the substance X given to the patient
y0: Level of the substance y before the operation
y1: Level of the substance y one day after the operation

4 General linear models

Question:

Analyze these data to find if the level of y one day after the operation depends on the treatment. Your model should account for the level of y before the operation, and for the age of the patient.

The solution should be similar to the "statistical methods" section of a scientific paper on the experiment, but may be written in more detail than in a biological paper. For example, you may suggest more than one analysis of the data. Also, comment on the fact that some patients randomized to treatment A were in fact given some treatment X. Does this affect your analysis?

Chapter 5

Nonparametric methods

Many statistical methods are parametric in the sense that a specified form of the distribution of the response variable is assumed. This assumption, of course, is open to criticism. In cases where the distribution is not known, or when the variable studied is not at least on an interval scale, nonparametric methods can be used. We will briefly summarize some commonly used nonparametric tests.

5.1 One sample: the sign test

The sign test can be used to test the hypothesis that a random sample of size n was taken from a population with median M_0. The hypotheses to be tested are

H_0 : $Md = M_0$
H_1 : $Md \neq M_0$ for a double-sided test
H_1 : $Md > M_0$ (or H_1: $Md < M_0$) for a single-sided test

If H_0 is true, then the probability that an observation is larger than M_0 is 0.5. It follows that the number of observations exceeding M_0 would be described by a binomial distribution with $p = 0.5$ and sample size n. This is used to calculate the p value of the test.

Example 5.1 *How seriously does the mother's alcohol consumption affect the development of children before birth? Jones et al (1974) studied this issue. Six women were found who had been chronic alcoholics during pregnancy. Their children were IQ tested at age 7, with the following*

5 Nonparametric methods

results:

$$78 \quad 102 \quad 64 \quad 84 \quad 100 \quad 70$$

The IQ scale is constructed such that the median is 100. Do the results indicate that the children to alcohol abusing mothers have a lower median IQ than 100?

The hypotheses to be tested are H_0: $Md = 100$ against the single-sided alternative H_1: $Md < 100$. Of the six children, four had an IQ below 100. The probability to obtain this result, or anything more extreme, given that H_0 is true, is obtained from the binomial distribution:

$$\begin{aligned} p \text{ value} &= P(x \geq 4 | p = 0.5) \\ &= p(x=4) + p(x=5) + p(x=6) \\ &= 0.234 + 0.094 + 0.016 = 0.344. \end{aligned}$$

Since the p value is larger than the conventional limit 0.05, the null hypothesis cannot be rejected.

It is interesting to note that a standard one-sample t test on these data gives $t = -2.31$ on 5 d.f. The single-sided 5% limit is 2.015 so we would reject H_0 if we could assume normality. This also illustrates the fact that the sign test is a rather weak test. If the underlying distribution is, in fact, normal, then the asymptotic relative efficiency of the sign test, as compared with the t test is $\frac{2}{\pi} \approx 0.64$. This means that the power of a sign test with $n = 100$ is approximately equal to the power of a t test with $n = 64$.

Sign test: To test H_0: $Md = M_0$ based on a sample of size n you calculate the test statistic

$$u = \text{Number of observations larger than } M_0.$$

For the single sided H_1: $Md < M_0$, the p value is given by $P(y \leq u)$.
For the single sided H_1: $Md > M_0$, the p value is given by $P(y \geq u)$.
For the double sided H_1: $Md \neq M_0$, the p value is given by
$$\begin{cases} 2P(y \leq u) & \text{if } u < n/2. \\ 2P(y \geq u) & \text{if } u > n/2, \\ \quad 1 & \text{if } u = n/2. \end{cases}$$
where in all cases $y \sim Bin(n, 0.5)$.

5.2 Two independent samples: the Wilcoxon-Mann-Whitney test

When the interest lies in comparing the "location" of two populations, the Wilcoxon-Mann-Whitney test can be used. It is an alternative to the two-sample t test when we have reasons to believe that the distributional assumptions are not fulfilled. It is based on the idea to replace the values of the variable by their ranks when the data are sorted. This is an idea that is used in other nonparametric procedures as well.

We assume that we have a sample of $n = n_1 + n_2$ observations from the two populations. The test is performed as follows:

1. Rank the n observations from the smallest to the largest. If many observations have the same value of y, they are given the average of the ranks.

2. Calculate the average rank of the observations in the two groups, say $\overline{R}_1$ and $\overline{R}_2$.

3. Calculate the test statistic

$$\chi^2 = \frac{12 \cdot n_1 \cdot n_2 \left(\overline{R}_1 - \overline{R}_2\right)^2}{n^2 (n+1)}.$$

For large samples, this can be compared with limits for the χ^2 distribution with 1 degree of freedom. For small samples (say, the smallest of n_1 and n_2 smaller than 10), tables of the exact sampling distribution of the test statistic, or a computer program, should be used. SAS (but not Minitab) can give exact p-values for this test.

Example 5.2 *The following data come from a clinical trial where the lysozyme levels in the gastric juice are compared for patients under two different treatments.*

5 Nonparametric methods

Lyozyme	Group	Rank	Lyozyme	Group	Rank
0.2	1	1.5	5.7	2	30.0
0.2	2	1.5	5.8	2	31.0
0.3	1	3.5	7.5	1	32.5
0.3	2	3.5	7.5	2	32.5
0.4	1	5.5	8.7	2	34.0
0.4	2	5.5	8.8	2	35.0
0.7	2	7.0	9.1	2	36.0
1.1	1	8.0	9.8	1	37.0
1.2	2	9.0	10.3	2	38.0
1.5	2	10.5	10.4	1	39.0
1.5	2	10.5	10.9	1	40.0
1.9	2	12.0	11.3	1	41.0
2.0	1	13.5	12.4	1	42.0
2.0	2	13.5	15.6	2	43.0
2.1	1	15.0	16.1	2	44.0
2.4	2	16.0	16.2	1	45.0
2.5	2	17.0	16.5	2	46.0
2.8	2	18.0	16.7	2	47.0
3.3	1	19.0	17.6	1	48.0
3.6	2	20.0	18.9	1	49.0
3.8	1	21.0	20.0	2	50.0
4.5	1	22.0	20.7	1	51.5
4.8	1	24.0	20.7	2	51.5
4.8	2	24.0	24.0	1	53.0
4.8	2	24.0	25.4	1	54.0
4.9	1	26.0	33.0	2	55.0
5.0	1	27.0	40.0	1	56.0
5.3	1	28.0	42.2	1	57.0
5.4	2	29.0	50.0	1	58.0
			60.0	1	59.0

The object is to compare the two treatments with respect to median lysozyme levels. The average ranks for the two treatments are 33.6552 and 26.4667, respectively, so the test statistic is

$$\chi^2 = \frac{12 \cdot 29 \cdot 30 \left(33.6552 - 24.4667\right)^2}{59^2 \cdot 60} = 2.58.$$

This is not significant; the 5% limit on 1 d.f. is 3.84.

The distribution of the test statistic is somewhat affected by the presence of ties, i.e. observations that share the same value of the variable. It is possible to compensate for the presence of ties in the following way. Say that there are t_1 ties at the first value, t_2 at the second, and so on. An adjustment to the test statistic above can be obtained by calculating

$$f = 1 - \frac{\sum t_i (t_i - 1)(t_i + 1)}{n(n-1)(n+1)}$$

$$\text{Adjusted } \chi^2 = \frac{\text{Unadjusted } \chi^2}{f}.$$

For observations without ties, there is no contribution to f. For the data in the example there are ties at 0.2 ($t_i = 2$), 0.3 ($t_i = 2$), 0.4 ($t_i = 2$),

1.5 ($t_i = 2$), 2.0 ($t_i = 2$), 4.8 ($t_i = 3$), 7.5 ($t_i = 2$) and 20.7 ($t_i = 2$). The adjustment f for these data is $f = 1 - \frac{66}{205.32} = 0.9997$ so the adjusted $\chi^2 = 2.58/0.9997 = 2.58$. For these data, the adjustment is negligible. A Minitab output of the test is as follows:

Mann-Whitney Test and CI: Group 1; Group 2

```
Group 1     N =  29      Median =         9.80
Group 2     N =  30      Median =         5.10
Point estimate for ETA1-ETA2 is          2.85
95.0 Percent CI for ETA1-ETA2 is  (-0.50;8.40)
W = 976.0
Test of ETA1 = ETA2  vs  ETA1 not = ETA2 is significant at 0.1097
The test is significant at 0.1096 (adjusted for ties)

Cannot reject at alpha = 0.05
```

Assume that $x_1,\ldots,x_{n_1}$ are observations from one population and $y_1,\ldots,y_{n_2}$ are observations from another population. We can test the hypotesis

$$H_0 : \text{"The medians are equal" against}$$
$$H_1 : \text{"The medians are different"}$$

by calculating the average rank $\overline{R}$ for each sample. If the samples are large, an approximate p value is obtained from the test statistic

$$\chi^2 = \frac{12 n_1 n_2 \left(\overline{R}_1 - \overline{R}_2\right)^2}{n^2 (n+1)}.$$

If the samples are small, a table of the exact distribution of the test statistic, or a computer package, should be used.

The asymptotic relative efficiency of the Mann-Whitney test, as compared with the two-sample t test, is $\frac{3}{\pi} \approx 0.955$ (Lehmann, 1975).

5.3 More than two samples: the Kruskal-Wallis test

The Kruskal-Wallis test is a generalization of the Wilcoxon-Mann-Whitney test to more than two treatments, where the experiment has a completely randomized design. It is an alternative to one-factor ANOVA.

The sample sizes for the a different groups are denoted with n_1, n_2, ..., n_a with $\sum n_i = n$. The test is performed similarly to the Wilcoxon

5 Nonparametric methods

test: the observations are ranked and the average ranks $\overline{R}_1, \overline{R}_2, \ldots \overline{R}_a$ are calculated for each treatment. The hypothesis of no treatment differences is tested using

$$\chi^2 = \frac{12 \sum n_i \left(\overline{R}_i - \frac{n+1}{2}\right)^2}{n(n+1)f}$$

which, under H_0, approximately follows a χ^2 distribution on $a-1$ degrees of freedom. The adjustment factor f in this expression is calculated as for the Wilcoxon test. The test is approximate and requires rather large samples. For small samples, tables of the exact distribution of the test statistic, or a computer package, should be used.

The asymptotic relative efficiency of the Kruskal-Wallis test, as compared with the standard F test, is $\frac{3}{\pi} \approx 0.955$ (Andrews, 1954).

Follow-up analyses after a significant over-all test could be done using pairwise Wilcoxon tests. However, there are more convenient alternatives that we will discuss in Section 5.5 (page 117).

Example 5.3 *Liss et al (1996) (see page 83) studied the urine production of rats which had been exposed to different contrast media. The change in urine production (before)−(after) was measured for each rat. An ANOVA, followed by residual diagnostics, suggests that the variances are unequal for the different media. Therefore, the data were analyzed with a Kruskal-Wallis test. The following results were obtained:*

Kruskal-Wallis Test: Diff versus Contrast medium

```
Kruskal-Wallis Test on Diff

Contrast    N    Median    Ave Rank         Z
Diatrizo    5   20.75000       51.8      3.22
Hexabrix    9    4.14400       23.9     -1.01
Isovist     9    1.52000       13.7     -3.02
Mannitol    9    7.51000       35.8      1.33
Omnipaqu   10    7.86650       37.1      1.70
Ringer      7    0.09700        6.0     -3.91
Ultravis    8    9.74000       40.1      2.04
Overall    57                  29.0

H = 38.89   DF = 6   P = 0.000
```

The test is highly significant (p=0.000) so we reject the hypothesis that the medians are equal for the different contrast media. The results can be followed up with pairwise comparisons between the media.

Table 5.1: Survival of skin patches for severely burnt patients

Patient	Compatibility		Sign
	Close	Poor	
1	37	29	+
2	19	13	+
3	57	15	+
4	93	26	+
5	16	11	+
6	23	18	+
7	20	26	−
8	63	43	+
9	29	18	+
10	60	42	+
11	18	19	−

5.4 Randomized block designs

5.4.1 Two samples: the sign test

The sign test, described on page 109, can also be used to analyze matched data. It is not very powerful, but it is a quick and handy tool to judge whether two treatments, applied in a matched-pairs design, differ significantly.

Example 5.4 *A sample of severely burnt patients were provided with temporary patches of skin from donors. It was suspected that close compatibility with respect to the HL-A antigen might delay the (inevitable) rejection of the patch. One piece of skin with close HL-A resemblance, and one piece of skin with poor compatibility, were applied on each patient. The data, summarized in Table 5.1 are number of days until the piece of skin was rejected.*

We want to test H_0: the survival time distributions are the same for both types of skin, against the single-sided alternative H_1: survival tends to be longer for pieces with close HL-A resemblance. The p value is calculated using the binomial distribution. For $n = 11$ patients we have observed $x = 9$ plus signs. The probability to observe this, or some more extreme case, is

$$p = P(x = 9) + P(x = 10) + P(x = 11)$$

where each probability is calculated using a binomial distribution with

5 Nonparametric methods

$p = 0.5$ and $n = 11$. We get the p value as

$$p = 0.02686 + 0.00537 + 0.00049 = 0.03272.$$

This is smaller than 0.05 so we reject the null hypothesis in favour of H_1. Survival time seems to be longer, on the average, for patches with a close HL-A resemblance.

5.4.2 More than two samples in randomized blocks: Friedman's test

If data have been collected as a randomized block design, the Friedman test can be used as an alternative to the traditional ANOVA. The test operates on the within-block ranks of the observations. It is assumed that there is no interaction between treatments and blocks. We also assume that there are a treatments and b blocks, and that no observations are missing so there are a total of $n = a \cdot b$ observations. For $b = 2$, this test is simply a large-sample approximation to the sign test.

The test procedure consists of the following steps:

1. Rank the observations within each block. Tied observations share the average of their ranks.

2. Calculate the sum of the ranks for each treatment, say $T_1, T_2, \ldots, T_a$.

3. Calculate the test statistic $S = \frac{12}{a \cdot b(a+1)} \sum_{i=1}^{a} T_i^2 - 3b(a+1)$. This is compared with limits for χ^2 on $(a-1)$ degrees of freedom. The test is approximate and requires large samples.

Example 5.5 *The data in Table 5.2 are measurements of microsomal epoxide hydrolase (mEH) activity in fish livers. Measurements were made using three different methods. Each method was used on tissues from the liver of each of five fish; thus the fish can be seen as a blocking factor.*

Data were analyzed using Friedman's method. The following results

Table 5.2: mEH activity in the livers of fish.

Fish	Method		
	A	B	C
1	3.8	12.3	17.7
2	6.2	15.3	17.1
3	12.8	13.6	20.5
4	11.4	12.3	13.5
5	12.5	12.9	13.9

were obtained:

Friedman Test: mEH versus Treatment; Fish

```
Friedman test for mEH by Treatmen blocked by Fish

S = 10.00   DF = 2   P = 0.007

                    Est      Sum of
Treatmen    N    Median     Ranks
A           5    10.333      5.0
B           5    12.900     10.0
C           5    16.367     15.0

Grand median  =  13.200
```

The result is significant ($p = 0.007$) indicating that the methods seem to give systematically different results.

5.5 Rank transformations: ANOVA on ranks

The Wilcoxon test, the Kruskal-Wallis test and the Friedman test are all examples of tests where the observed data have been transformed into ranks for analysis. A more direct way of analyzing such data is to proceed as follows. Rank the data (within each block, if appropriate). You then make a *rank transformation*. Use the rank numbers as input to a procedure for general linear models, for example, the GLM procedure in SAS. The resulting output will include the usual F test. This gives (in large samples) a good approximation to the correct p-values. Thus, making an ANOVA on the rank numbers can be used as an alternative to the χ^2 approximation used above. The advantage with this approach is that it is more general. It can be used for many types of experimental designs (with two samples, with a samples, in block designs). In addition, the post hoc tests available in Proc GLM can be used, with or without adjustment for multiplicity to perform follow-up analyses after

5 Nonparametric methods

a significant result.

In SAS, the procedure Proc RANK calculates the rank numbers from a given dataset and produces an output dataset that Proc GLM can use. Alternatively, the ranks can be calculated in software such as Excel.

Example 5.6 *Liss et al (1996; see page 83) studied how contrast media affect the urine production of mice. We have analyzed these data using a Kruskal-Wallis test (Page 114). A similar analysis can be performed using ANOVA on ranks. First, ask Proc RANK to calculate ranks and store them in the data set Ranks under the variable name Rank:*

```
PROC RANK data=Liss out=ranks;
VAR diff;
RANKS Rank;
RUN;
```

Then make en ANOVA on the rank values:

```
PROC GLM data=ranks;
CLASS medium;
MODEL rank = medium;
LSMEANS medium /ADJUST=bon ;
RUN;
```

Output:

```
Dependent Variable: Rank    Rank for Variable Diff
```

Source	DF	Sum of Squares	Mean Square	F Value	Pr > F
Model	6	10712.98056	1785.49676	18.94	<.0001
Error	50	4714.51944	94.29039		
Corrected Total	56	15427.50000			

In the program we also asked for pairwise comparisons between the treatments using Bonferroni adjustment of p-values. This gives the following results:

```
Least Squares Means for effect Medium
  Pr > |t| for H0: LSMean(i)=LSMean(j)
```

Medium	Rank LSMEAN	LSMEAN Number
Diatrizoate	51.8000000	1
Hexabrix	23.8888889	2
Isovist	13.6666667	3
Mannitol	35.7777778	4
Omnipaque	37.1000000	5
Ringer	6.0000000	6
Ultravist	40.1250000	7

i/j	1	2	3	4	5	6	7
1		<.0001	<.0001	0.0990	0.1675	<.0001	0.8394
2	<.0001		0.6309	0.2585	0.0983	0.0129	0.0248
3	<.0001	0.6309		0.0003	<.0001	1.0000	<.0001
4	0.0990	0.2585	0.0003		1.0000	<.0001	1.0000
5	0.1675	0.0983	<.0001	1.0000		<.0001	1.0000
6	<.0001	0.0129	1.0000	<.0001	<.0001		<.0001
7	0.8394	0.0248	<.0001	1.0000	1.0000	<.0001	

The F test gives a good approximation to the Kruskal-Wallis statistic. In addition, this approach has the advantage that pairwise comparisons (and other contrasts) can be tested in a simple way.

5.6 Randomization tests

Randomization tests are nonparametric tests that regard the observed data as given. The p value is obtained by considering how many rearrangements of the data that exist that would lead to a larger value of the chosen statistic. This is compared with the total number of rearrangements of the data that are possible. The principle underlying randomization tests is best explained by an example.

5.6.1 Fisher's exact test

Example 5.7 *An experiment has been made to compare two species of Acacia trees with respect to their resistance to being occupied by ants (Sokal and Rohlf, 1982). A number of trees of each variety were cleaned from insects. Then, ant colonies were placed such that they could choose which tree to occupy. The result is summarized in Table 5.3.*

Table 5.3: Number of trees of two species occupied by ants

Species	Occupied	Not occupied	Total
A	2	13	15
B	10	3	13
Total	12	16	28

We want to test the hypothesis that the probability of occupation is the same for both species: H_0: $p_1 = p_2$. The samples are so small that a a z test, or a chi-square test, (see chapter 6) would give unreliable results.

A randomization test of H_0 proceeds as follows. We regard de observed marginal totals as fixed. For given values of the totals, we will find if it would have been possible to find tables that are more extreme

5 Nonparametric methods

than the observed one, in terms of our alternative hypothesis. It turns out that the two coss-tables presented in Table 5.4 are the only ones that would have been "more extreme".

Table 5.4: Two cross tables that are more extreme than the one we observed.

Species	Not occu-pied	Occu-pied	Total	Not occu-pied	Occu-pied	Total
A	1	14	15	0	15	15
B	11	2	13	12	1	13
Total	12	16	28	12	16	28

Thus there are three tables that are equally or more extreme than the one we got. To examine whether or not H_0 should be rejected, we need to calculate the probability of this result, if H_0 is true.

A cross-table with 2 rows and 2 columns can be written as

Species			Sum
A	a	b	n
B	c	d	m
Sum	S	F	N

The probability that a of the S trees that were invaded, belongs to species A, given that H_0 is true, can be calculated using the hypergeometric distribution:

$$P(a) = \frac{\binom{S}{a}\binom{N-S}{n-a}}{\binom{N}{n}}.$$

For the three cross-tables we study we get the following probabilities. The sum of these probabilities gives our p value:

$$p = 0.000987 + 0.000038 + 0.000000 = 0.001026$$

The result is significant at least at the 1% level. It seems that the ants prefer to occupy trees of species A.

5.6.2 Other applications of randomization tests

Similar ideas as in Fisher's exact test can be applied also in other situations.

Example 5.8 *Assume that we want to compare two treatments A_1 and*

A_2 in a completely randomized design. The following data are obtained:

	A_1	A_2
	5.0	1.0
	7.0	3.0
	8.0	4.0
	10.0	6.0
Mean	7.5	3.5

A randomization test of the hypothesis that the two treatments have the same population mean is made as follows. The difference in mean value between the treatments is $7.5 - 3.5 = 4$. There are

$$\frac{N!}{n_1! n_2!} = \frac{8!}{4!4!} = 70$$

ways to randomly distribute eight numbers into two groups of size four. Of these, there are only two ways in which the difference between the sample means is 4 or larger:

	A_1	A_2	A_1	A_2
	5	1	6	1
	7	3	7	3
	8	4	8	4
	10	6	10	5
Mean	7.5	3.5	7.75	3.25

This means that the probability to obtain a difference in sample mean of 4 or larger, if we randomly distribute the eight values into two groups, is

$$p \text{ value} = \frac{2}{70} \approx 0.029.$$

Since the p value is smaller than 0.05, we would reject the hypothesis that the population means are equal.

Randomization tests are very general and flexible. They can be used on most types of statistics (mean, median, variance...). Also, randomization tests are rather powerful: the asymptotic power of a randomization test is equal to that of the corresponding parametric test. The principle of a randomization test is as follows:

1. Decide on a key parameter to test.

2. Calculate the number of possible samples. For an ANOVA situa-

tion with a treatments the number of possible samples is

$$M = \frac{N!}{n_1! n_2! \ldots n_a!}.$$

3. For each sample, calculate an estimate of the key parameter.
4. Record how many of the samples (say m) that have an estimate of the key parameter equal to or larger than the one we obtained.
5. The p value of the single sided test is $\frac{m}{M}$.

But randomization tests have drawbacks too. Consider a case where we have three treatments, each with 15 observations. The number of ways the data can be distributed is then

$$\frac{45!}{15! 15! 15!} = 5.34 \cdot 10^{19}.$$

Even a high-speed computer would be hard put to calculate the key parameter for each of these samples. If the sample sizes are even larger, the computational problems may become insurmountable.

As an alternative, an approximate solution based on re-sampling has been suggested. The idea is to take a random "sample of samples" among the large number of possible samples, calculate the key statistic only in these, and in this way get a reasonable estimate of the p value. If we take, for example, one million samples, the estimate of the p value will be rather good, while at the same time the computational burden is tractable.

Note that the principle of randomization tests can be used even on many parametric test statistics, such as t tests and F tests, thus making even those tests non-parametric. Randomization tests are available in the nonparametric ANOVA procedure in SAS, as a "wrapper" for any SAS procedure, and in some R packages.

5.7 Some comments on nonparametric tests

The nonparametric tests discussed in this chapter share the property that they do not require that the data can be modeled using e.g. a normal distribution. Note, however, that other assumptions are still required for the tests to be valid. The most important assumption is that the observations are independent of each other. One way to achieve this is through random sampling.

A common mistake is to apply simple nonparametric methods in situations where the data include several observations made on the same individuals, for example at different time points. In such data, it can be assumed that observations within individual are dependent. Then, the analysis should first of all focus on modeling this dependence. Possible problems with the normality assumption are then of secondary importance.

Also note that parametric methods like ANOVA are rather robust to deviations from normality, in particular if the samples are large and the experiment is balanced. This means that modest deviations from normality have little effect on the p values. In many cases, transformation of the data using e.g. a log transformation may improve the analysis. If this is not possible, an analysis using the correct model, but using re-sampling methods (section 5.6.2) to obtain correct p values, may be a good choice.

5 Nonparametric methods

5.8 Exercises

Exercise 5.1 *High levels of triglycerides in blood serum may be related to the risk of myocardial infarction. A study has been made to investigate whether training may affect the level of triglycerides. Seven male volunteers underwent a training program lasting for ten weeks. The level of triglycerides was measured before and after the program. Results:*

Person	Before	After	Change
1	0.87	0.57	0.30
2	1.13	1.03	0.10
3	3.14	1.47	1.67
4	2.14	1.43	0.71
5	2.98	1.20	1.78
6	1.18	1.09	0.09
7	1.60	1.51	0.09

A. *All values of "Change" are positive. Make a sign test to test whether the population median change may be equal to zero.*

B. *As a comparison, carry out a matched t test to test whether the population mean change may be zero. Make a verbal conclusion and give the assumptions underlying the analyses.*

Exercise 5.2 *Darwin (1876) tried to show that plants that had been multiplied through cross-fertilization were more "agile" than plants multiplied through self-fertilization. Darwin raised plants of both types. The plants were planted in pairs, with one cross-fertilized and one self-fertilized plant in each pair. Thus, the design was a matched pairs design or, which is equivalent, a randomized block design with "pair" as block. The following table gives the length of each plant:*

Pair	1	2	3	4	5	6	7
Cross	23.5	12.0	21.0	22.0	19.1	21.5	22.1
Self	17.4	20.4	20.0	20.0	18.4	18.6	18.6

Pair	8	9	10	11	12	13	14	15
Cross	20.4	18.3	21.6	23.3	21.0	22.1	23.0	12.0
Self	15.3	16.5	18.0	16.3	18.0	12.8	15.5	18.0

We have used these data earlier to illustrate t tests. We will here analyze the same data nonparametrically, without assuming normality. This may be relevant since the data seem to include a few "outliers". The hypothesis of interest is whether cross-fertilized plants tend to be longer than self-fertilized plants.

Analyze these data using some nonparametric test that does not assume normality. State the assumptions underlying the analysis and indicate which hypothesis is being tested.

Exercise 5.3 The reaction times of a number of volunteers were measured after the persons had taken one of two drugs, A and B (data from (McClave and Sincitch, 2009). In one of the groups, one person dropped out. It is of interest to compare the drugs with respect to average reaction times. Data are as follows:

Drug A		Drug B	
Time	Rank	Time	Rank
1.96	4	2.11	6
2.24	7	2.43	9
1.71	2	2.07	5
2.41	8	2.71	11
1.62	1	2.50	10
1.93	3	2.84	12
		2.88	13

Perform a non-parametric test comparing the reaction times for the two drugs. Even if the samples are not very large, use the χ^2 approximation to decide if the result is significant.

Chapter 6

Analysis of frequency data

In this chapter we will study the analysis of data where the response variable is categorical (non-numeric). Some examples of categorical variables are color, blood group, and "Alive/dead". Some of the methods discussed can also be used on grouped numerical data.

Categorical data are often summarized as tables. We observe how many times we have obtained each value of the response variable. It is these observed frequencies that can help us to draw conclusions about the population from which the data were drawn.

We will start by looking at the situation where we have one single sample, and where the purpose of the experiment is to draw conclusions about the probability distribution of the population. Later in the chapter we will compare several samples where the data can be summarized as a cross-tabulation.

6.1 Test of model fit

6.1.1 Testing model fit: the distribution is completely specified

Example 6.1 *A genetic trial with a certain variety of flowers has resulted in 120 violet flowers with green pistils; 48 violet flowers with red pistils; 36 red flowers with green pistils; and 13 red flowers with red pistils. According to genetic theory, the number of flowers with these properties should occur according to the relation 9:3:3:1. We will investigate if the observed numbers, which will be denoted with O_i, are consistent with the genetic theory.*

6 Analysis of frequency data

The genetic theory can be stated as the null hypothesis:

$$H_0: p_1 = 9/16, \ p_2 = 3/16, \ p_3 = 3/16 \text{ and } p_4 = 1/16.$$

As a first step, we can calculate how many flowers we would get of each type if our sample data had been exactly according to the theory. We have 217 flowers in total. If the probability that a flower is violet with a green pistil is 9/16, we would expect to get $9/16 \cdot 217 = 122.06$ flowers of this kind. In the same way, we would expect $3/16 \cdot 217 = 40.69$ violet flowers with red pistil, $3/16 \cdot 217 = 40.69$ red flowers with green pistil, and $1/16 \cdot 217 = 13.56$ red flowers with red pistil. These so called *expected frequencies* were calculated as

$$E_i = p_i \cdot n,$$

where p_i is the probability to get a flower of type i, and n is the total number of flowers. We keep the decimals, in spite of the fact that E_i are called "frequencies". The results are summarized in Table 6.1.

Table 6.1: Observed and expected number of flowers of different types in a genetic trial.

	\multicolumn{4}{c}{Flower of type}			
	1	2	3	4	Total
O_i	120	48	36	13	217
E_i	122.06	40.69	40.69	13.56	217

If the observed frequencies are close to the expected frequencies, we would not reject H_0, while large differences between O_i and E_i would indicate that H_0 is false. To test the hypothesis we calculate

$$\chi^2 = \sum \frac{(O_i - E_i)^2}{E_i}.$$

If H_0 is true, this statistic approximately follows a chi-square distribution. χ^2 includes k terms (which is the number of types of the flowers). It can be shown that χ^2 has $k-1$ degrees of freedom. In our example, $k = 4$ so chi-square will have $4 - 1 = 3$ degrees of freedom. We calculate

$$\begin{aligned}\chi^2 &= \frac{(120 - 122.06)^2}{122.06} + \frac{(48 - 40.69)^2}{40.69} + \frac{(36 - 40.69)^2}{40.69} + \frac{(13 - 13.56)^2}{13.56} \\ &= 1.91.\end{aligned}$$

The 5% limit for χ^2 on 3 degrees of freedom is 7.815. Our observed value

1.91 is smaller than the limit 7.815, so the result is not significant. H_0 cannot be rejected.

Note that H_0 is rejected only if we get an unusually large value of χ^2. A small value of χ^2 means that the observed frequencies are very close to the expected frequencies. Also note that this test is only approximate: it requires rather large samples. For the χ^2 approximation to be reasonable, all expected frequencies should be larger than 5.

To test hypotheses of type $p_i = a_i$, where the numbers a_i are given, we first calculate expected frequencies as

$$E_i = np_i.$$

Then we calculate

$$\chi^2 = \sum \frac{(O_i - E_i)^2}{E_i}.$$

The result is compared with limits for χ^2 on $k-1$ degrees of freedom. Assumptions: We have a random sample from some population. The expected frequencies E_i are at least 5.

6.1.2 Merging categories

The χ^2 test is approximate: it works well only if the samples are "large". A commonly used rule of thumb is that all expected frequencies should be larger than 5. If you need to analyze data where some expected frequencies are small, it is sometimes possible to merge categories such that all expected frequencies are large enough.

Example 6.2 *During the second world war, the Germans launched V2 bombs against London. If the bombs could be aimed at specific targets, some areas would get hit by more bombs, but if the bombs could not be aimed, they would be spread randomly over London. For 576 square areas in London it was noted how many V2 bombs that had hit the square. If the bombs could not be aimed, the number of bomb hits per square would follow a Poisson distribution. Let us first investigate if the distribution of number of hits resembles a Poisson distribution with mean value $\lambda = 1$. Table 6.2 summarizes the data (the observed numbers O_i), the probability p_i to have a certain number of hits (from the Poisson distribution table) and the expected numbers $E_i = np_i$.*

In Table 6.2 we note that the expected numbers are rather small for $x \geq 4$. Therefore, we merge classes to get Table 6.3.

6 Analysis of frequency data

Table 6.2: Number of V2 bombs per square in London during World war II. Data from Clarke (1946).

Number of hits x	O_i	$\widehat{p}_i$	$E_i = n\widehat{p}_i$
0	229	0.368	212.0
1	211	0.368	212.0
2	93	0.184	106.0
3	35	0.061	35.1
4	7	0.015	8.6
5	0	0.003	1.7
6	0	0.001	0.6
7	1	0.000	0.0
Total	$n = 576$	1.000	576.0

Table 6.3: Clarke's data on V2 bombs in London. The classes from $x = 4$ and upwards have been merged.

Number of hits x	O_i	$\widehat{p}_i$	E_i
0	229	0.368	212.0
1	211	0.368	212.0
2	93	0.184	106.0
3	35	0.061	35.1
≥ 4	8	0.019	10.9
Total	576	1.000	576.0

We calculate χ^2 in the same way as before:

$$\begin{aligned}\chi^2 &= \sum \frac{(O_{ij} - E_{ij})^2}{E_{ij}} = \frac{(229 - 212.0)^2}{212.0} + \frac{(211 - 212.0)^2}{212.0} + \\ &\quad \frac{(93 - 106.0)^2}{106.0} + \frac{(35 - 35.1)^2}{35.1} + \frac{(8 - 10.9)^2}{10.9} \\ &= 3.734.\end{aligned}$$

This can be compared with the 5% limit for χ^2 on $k - 1 = 5 - 1 = 4$ degrees of freedom; the limit is 9.488. The result is not significant; we cannot reject the hypothesis that the data are from a Poisson distribution with mean value $\lambda = 1$.

6.1.3 Test of model fit: the parameters are unknown

One could ask why we used the mean value $\lambda = 1$ for the Poisson distribution in Example 6.2. It would be sensible to simply ask the question

Test of model fit

"Can the data come from a Poisson distribution?", without stating the mean value. As an estimator of the mean value we can calculate the sample mean and use this as an estimated value of λ. For our data we would get

$$\widehat{\lambda} = \bar{x} = \frac{\sum x_i O_i}{n} = \frac{0 \cdot 229 + 1 \cdot 211 + 2 \cdot 93 + 3 \cdot 35 + 4 \cdot 7 + 7 \cdot 1}{576} = 0.93.$$

To test if the data can come from a Poisson distribution with mean 0.93 we do as above: we calculate $\widehat{p}_i$ (this time using a calculator or a computer). Then we calculate $E_i = n\widehat{p}_i$, and compute χ^2 in the same way as above. Table 6.4 summarizes the observed and expected frequencies, after merging some classes.

Table 6.4: Observed and expected numbers for Clarkes data on V2 bombs in London. The expected numbers were calculated from a Poisson distribution with $\widehat{\lambda} = 0.93$.

Number of hits x	O_i	$\widehat{p}_i$	E_i
0	229	0.395	227.3
1	211	0.367	211.4
2	93	0.171	98.3
3	35	0.053	30.5
≥ 4	8	0.015	8.6
Total	576	1.000	576.0

The value of χ^2 is 1.26. As for degrees of freedom ν, the following rule applies:

$$\nu = k - 1 - \text{the number of parameters estimated}.$$

For the test, we have estimated one parameter: the mean value λ in the Poisson distribution. Thus we get $5 - 1 - 1 = 3$ degrees of freedom. The 5% limit for χ^2 on $\nu = 3$ d.f. is 7.815, which is much larger than our observed value 1.26. The test is far from significant; the data agree fairly well with a Poisson distribution.

The Englishmen were quite happy about this result during the war. The fact that the data agree well with a Poisson distribution means that the bombs hit randomly: they could not be aimed at specific targets, such as Buckingham Palace.

6 Analysis of frequency data

> To test the hypothesis that a set of data comes from a distribution that is specified, apart from r unknown parameters, we estimate the parameters from the data. Then, expected frequencies are calculated as
> $$E_i = np_i$$
> where the probabilities p_i are calculated from the distribution in question, using the estimated parameter values. Then we calculate
> $$\chi^2 = \sum \frac{(O_i - E_i)^2}{E_i}.$$
> The result is compared with limits for χ^2 on $k - 1 - r$ degrees of freedom, where r is the number of parameters estimated. Assumptions: We have a random sample from some population. The expected frequencies E_i are at least 5.

6.2 Testing homogeneity

A test of homogeneity is used to test the hypothesis that r samples are taken from the same discrete unspecified distribution. The data can be summarized as a crosstable with r rows and c columns. Each sample is on one row and the c columns are the different values of the variable; see Table 6.5.

Table 6.5: The structure of the data for testing homogeneity.

Sample	Category 1	2	...	c	Sum
1	O_{11}	O_{12}	...	O_{1c}	$n_{1.}$
2	O_{21}	O_{22}		O_{2c}	$n_{2.}$
$\vdots$	$\vdots$	$\vdots$	$\ddots$	$\vdots$	$\vdots$
r	O_{r1}	O_{r2}	...	O_{rc}	$n_{r.}$
Sum	$n_{.1}$	$n_{.2}$	...	$n_{.c}$	n

Example 6.3 *Pupae from each of three species of insects that may damage the collections in museums were exposed to reduced air pressure (vacuum) for twelve hours. The purpose was to investigate if this method can be used as an ecological method for pest control e.g. in books. Thirty pupae from each species were used. After twelve hours, the number of surviving insects was counted. The results are given in Table 6.6.*

Table 6.6: Number of surviving and dead insects of different species after twelve hours treatment in vacuum.

	Alive	Dead	Total
A. museorum	$O_{11} = 10$	$O_{12} = 20$	30
P. tectus	$O_{21} = 16$	$O_{22} = 14$	30
A. woodroffei	$O_{31} = 22$	$O_{32} = 8$	30
Total	48	42	90

In this analysis, we will investigate whether there are any differences between the species in the probability of survival. The null hypothesis is

$$H_0: p_{1.} = p_{2.} = p_{3.}$$

where museorum is denoted with 1, tectus with 2 and woodroffei with 3. The dots indicate averaging over the second index. Is the difference between these proportions so large that it cannot be attributed to chance?

If, in the population, $p_{1.} = p_{2.} = p_{3.} = p$ we would expect $E_{11} = n_{1.} \cdot p$ surviving A. museorum, $E_{21} = n_{2.} \cdot p$ surviving P. tectus and $E_{31} = n_{3.} \cdot p$ surviving A. woodroffei. We estimate p as $\widehat{p} = 48/90$, i.e. as the proportion of survivors in the whole sample. Then, E_{11} can be calculated as $30 \cdot 48/90 = 16$. In the same way, the expected number in row number i, column number j in the table can be calculated as

$$E_{ij} = \frac{(\text{Sum in row } i)(\text{sum in column } j)}{n} = \frac{n_{i.} n_{.j}}{n}.$$

The expected frequencies, calculated in this way, are summarized in Table 6.7.

Table 6.7: Expected frequencies for data in Example 6.3.

	Alive	Dead	Total
A. museorum	$E_{11} = 16$	$E_{12} = 14$	30
P. tectus	$E_{21} = 16$	$E_{22} = 14$	30
A. woodroffei	$E_{31} = 16$	$E_{32} = 14$	30
Total	48	42	90

As a measure of the difference between the observed frequencies O_{ij} and the expected frequencies E_{ij} we calculate

$$\chi^2 = \sum \frac{(O_{ij} - E_{ij})^2}{E_{ij}}.$$

6 Analysis of frequency data

The distribution of this statistic can, under certain assumptions, be approximated with a χ^2 distribution. The degrees of freedom of χ^2 is equal to

$$(\text{number of rows} - 1) \cdot (\text{number of columns} - 1) = (r-1)(c-1).$$

Our table has three rows and two columns, so we get $(3-1)(2-1) = 2$ degrees of freedom. The table of χ^2 with 2 degrees of freedom gives the 5% limit as 5.991, the 1% limit as 9.210 and the 0.1% limit as 13.816. Our data gives:

$$\begin{aligned}
\chi^2 &= \sum \frac{(O_{ij} - E_{ij})^2}{E_{ij}} \\
&= \frac{(10-16)^2}{16} + \frac{(20-14)^2}{14} + \frac{(16-16)^2}{16} + \frac{(14-14)^2}{14} \\
&\quad + \frac{(22-16)^2}{16} + \frac{(8-14)^2}{14} \\
&= 9.64.
\end{aligned}$$

This is significant at least at the 1% level. It seems that the three species have different survival probability.

> To test the hypothesis that two (or more) samples come from the same discrete distribution we calculate expected frequencies as
>
> $$E_{ij} = \frac{(\text{Sum in row } i)(\text{Sum in column } j)}{n}.$$
>
> Then we calculate
>
> $$\chi^2 = \sum \frac{(O_{ij} - E_{ij})^2}{E_{ij}}.$$
>
> This is compared with limits for χ^2 on $(r-1)(c-1)$ d.f. Assumptions: We have a random sample from each population. The expected frequencies E_i are at least 5.

6.3 Testing independence

Example 6.4 *In a study of the relation between snoring and heart problems (Norton and Dunn, 1985), data were collected for 2484 patients. One part of the information gathered was whether the patient had heart problems, and whether the person used to snore while asleep. The results* *are summarized in Table 6.8. An interesting question is whether there is any relation between snoring and heart problems.*

Table 6.8: Number of persons with and without heart problems distributed on snoring category.

Heart problems	Snores				
	Never	Sometimes	Often	Always	Total
Yes	24	35	21	30	110
No	1355	603	192	224	2374
	1379	638	213	254	2484

A model for the data can be formulated as follows. Our sample comes from a population where the different probabilities can be summarized as a crosstable (Table 6.9), The observed frequencies O_{ij} correspond to probabilities (see Table 6.10).

Table 6.9: The structure of data for a test of independence.

	Column category				
Row category	1	2	...	k	Sum
1	O_{11}	O_{12}	...	O_{1k}	$n_1.$
2	O_{21}	O_{22}		O_{2k}	$n_2.$
$\vdots$	$\vdots$		$\ddots$		$\vdots$
r	O_{r1}	O_{r2}		O_{rk}	$n_r.$
Sum	$n._1$	$n._2$		$n._k$	n

Table 6.10: The population in a test of independence.

	Column category				
Row category	1	2	...	k	Sum
1	p_{11}	p_{12}	...	p_{1k}	$p_1.$
2	p_{21}	p_{22}		p_{2k}	$p_2.$
$\vdots$	$\vdots$		$\ddots$		$\vdots$
r	p_{r1}	p_{r2}		p_{rk}	$p_r.$
Sum	$p._1$	$p._2$		$p._k$	1

The hypothesis to be tested is

H_0: The row and column categories are independent.

6 Analysis of frequency data

If the null hypothesis is true it should hold that

$$p_{ij} = p_{i.}p_{.j} \text{ for all } i \text{ and } j.$$

Thus the null hypothesis can be written as

$$H_0: p_{ij} = p_{i.}p_{.j} \text{ for all } i \text{ and } j.$$

Assuming that H_0 is true, the expected numbers E_{ij} can be calculated as

$$E_{ij} = n \cdot p_{i.} \cdot p_{.j}.$$

However, the probabilities of type $p_{i.}$ are unknown. We can estimate these from data as

$$\widehat{p}_{i.} = \frac{n_{i.}}{n}$$
$$\widehat{p}_{.j} = \frac{n_{.j}}{n}$$

Now the expected frequencies can be calculated as

$$E_{ij} = n\widehat{p}_{i.}\widehat{p}_{.j} = n\frac{n_{i.}}{n}\frac{n_{.j}}{n} = \frac{n_{i.}n_{.j}}{n}.$$

This, in fact, is the same as

$$E_{ij} = \frac{(\text{sum in row } i) \cdot (\text{sum in column } j)}{n} = \frac{n_{i.}n_{.j}}{n}.$$

The null hypothesis is tested by calculating

$$\chi^2 = \sum \frac{(O_{ij} - E_{ij})^2}{E_{ij}}.$$

The result is compared with percentage points for a χ^2 distribution with $(r-1)(c-1)$ degrees of freedom.

For the data in Example 6.4, the calculations are made as follows. We calculate the expected frequencies. For example, $E_{11} = \frac{110 \cdot 1379}{2484} = 61.07$. The observed and expected frequencies are summarized in Table 6.11. The expected frequencies are given within parenthesis.

Table 6.11: Observed and expected frequencies for data on the relation between snoring and heart problems.

	Snores				
Heart problems	Never	Sometimes	Often	Always	Total
Yes	24	35	21	30	110
	(61.07)	(28.25)	(9.43)	(11.25)	
No	1355	603	192	224	2374
	(1317.93)	(609.75)	(203.57)	(242.75)	
	1379	638	213	254	2484

Now, χ^2 can be calculated as

$$\chi^2 = \sum \frac{(O_{ij} - E_{ij})^2}{E_{ij}}$$

$$= \frac{(24 - 61.07)^2}{61.07} + \frac{(35 - 28.25)^2}{28.25} + \frac{(21 - 9.43)^2}{9.43} + \frac{(30 - 11.25)^2}{11.25}$$

$$+ \frac{(1355 - 1317.93)^2}{1317.93} + \frac{(603 - 609.75)^2}{609.75} + \frac{(192 - 203.57)^2}{203.57}$$

$$+ \frac{(224 - 242.75)^2}{242.75} = 72.78.$$

Here, χ^2 has $(r-1)(c-1) = (2-1)(4-1) = 3$ degrees of freedom. The limits are 7.815 (5%); 11.345 (1%) and 16.286 (0.1%). Our observed value of χ^2 is larger than all these limits. The result is strongly significant and we can reject H_0. There seems to be a relation between snoring and heart problems.

To test the hypothesis of independence between the variables in an $r \cdot c$ cross table we calculate expected frequencies

$$E_{ij} = \frac{(\text{Sum in row } i)(\text{Sum in column } j)}{n}.$$

Then we calculate

$$\chi^2 = \sum \frac{(O_{ij} - E_{ij})^2}{E_{ij}}.$$

This is compared with limits for χ^2 on $(r-1)(c-1)$ df. Assumptions: We have a random sample from some population. The expected frequencies E_i are at least 5.

This analysis can also be done using programs like SAS or Minitab. If the data are in the form of raw data, with one row for each person,

6 Analysis of frequency data

Minitab can compile the table and at the same time do the test. A χ^2 test can also be performed if you already have a table to feed into the program.

Tabulated statistics: Heart problems; Snoring

```
Rows: Heart problems   Columns: Snoring

              1        2        3        4       All

   No      1355      603      192      224      2374
          1317.9    609.7    203.6    242.8    2374.0
           1.043    0.075    0.657    1.449       *

   Yes       24       35       21       30       110
            61.1     28.3      9.4     11.2     110.0
          22.499    1.611   14.186   31.262       *

   All     1379      638      213      254      2484
          1379.0    638.0    213.0    254.0    2484.0
             *        *        *        *         *

Cell Contents:         Count
                       Expected count
                       Contribution to Chi-square

Pearson Chi-Square = 72.782; DF = 3; P-Value = 0.000
```

We have found a significant relation between snoring and heart problems. We can follow up the analysis by looking at individual terms in the calculation of χ^2. These are printed in the Minitab output. Large contributions to χ^2 are from persons with heart problems who snore seldom, and from persons without heart problems who always snore. In the raw data, there are 24 persons in the first category while the expected number is 61.07. There are 30 persons in the second category; the expected number is 11.25. Thus, persons with heart problems are under-represented in the category "never snores", and over-represented in the category "always snores". This indicates the direction of the relation between snoring and heart problems.

6.4 Some comments on χ^2 tests

The χ^2 test is approximate. We have mentioned that the form of χ^2 tests presented here are only approximate. The tests work only if the samples are "large". We have given the rule of thumb that all expected numbers should be larger than 5. Other textbooks have other rules. For example, the Minitab manual suggests that the test is doubtful if more than 20% of the cells have expected numbers smaller than 5, in particular if these cells contribute much to the total χ^2 value. If you need to analyze a crosstable where many expected numbers are small, you can merge categories and get a smaller table. Alternatively, you can use randomization tests that are available in some computer packages.

Never make a χ^2 test on percentages. The χ^2 tests we have presented here are based on the assumption that observed numbers are analyzed. The tests do not work if the analysis is made on percentages.

Tests of independence vs. tests of homogeneity. The two examples above may cause some confusion. We have presented two different tests and two partly different situations, but in the end the tests were the same. One might ask why. We can note one difference between the χ^2 test on page 132 and the test on page 134. In the first case, the sample sizes were fixed: we decided beforehand how many insects of the two species we would use in the experiment. Some of the marginal sums were fixed. In the second example (page 134) the total number of persons with and without heart problems is random: we did not decide to sample a fixed number of persons of each category. In the first case, we have two samples of pupae, one sample of each species. In the second example we have one sample of individuals, who can be classified as snorers/non-snorers and as patients with/without heart problems. The situation where some margins are fixed is called a test of homogeneity, whereas the situation where the margins are random is called a test of independence. However, we can note that the test is performed in the same way in both cases. In the simple examples we will treat here, the calculation of χ^2 is the same, and the number of degrees of freedom is not affected whether the margins are fixed or random.

Relation to the z test In the $2 \cdot 2$ cross tables, a z test can be used to test if two probabilities are equal. How is the z test of the hypothesis $H_0: p_1 = p_2$ related to the χ^2-test? Let us look at the data for the species A. museorum and P. tectus on page 132. For A. museorum, 10 out of 30 survived while 16 out of 30 survived for P. tectus. To test H_0: $p_1 = p_2$ we can make a z test. We use that $\widehat{p}_0 = 26/60$. The z test is

$$z = \frac{\widehat{p}_1 - \widehat{p}_2}{\sqrt{\widehat{p}_0(1-\widehat{p}_0)\left(\frac{1}{n_1}+\frac{1}{n_2}\right)}} = \frac{10/30 - 16/30}{\sqrt{26/60(1-26/60)\left(\frac{1}{30}+\frac{1}{30}\right)}} = -1.563.$$

The same hypothesis tested with a χ^2 test gives the following Minitab output:

6 Analysis of frequency data

Tabulated statistics: Species; Result

```
Using frequencies in f

Rows: Species   Columns: Result

           Alive   Dead   All

Museorum      10     20    30
Tectus        16     14    30
All           26     34    60

Cell Contents:      Count

Pearson Chi-Square = 2.443; DF = 1; P-Value = 0.118
```

It turns out that $z^2 = (-1.563)^2 = 2.44$, which is the value of χ^2 presented by Minitab. This holds in general: a χ^2 test in a $2 \cdot 2$ cross table gives the same answer as the square of the z test statistic. And the limits are also related, The 5% limit for z is 1.96. We square this to obtain $1.96^2 = 3.84$ which is the 5% limit for χ^2 on 1 degree of freedom. The conclusion is that the z test and the χ^2 test are equivalent for $2 \cdot 2$ cross tables.

Continuity correction As noted above, the chi-square test is approximate: we approximate a discrete distribution based on observed frequencies with a continuous distribution. Yates (1934) showed that the approximation can be improved by using as test statistic

$$\chi^2 = \sum \frac{(|O_{ij} - E_{ij}| - 0.5)^2}{E_{ij}}$$

i.e. by subtracting 0.5 in the numerator. This is used in some computer packages. Our recommendation, if sample sizes are small, is to use randomization tests.

In this chapter we have discussed analysis of rwo-dimensional contingency tables. Larger tables can be analyzed using log-linear models; see Page 157.

6.5 Exercises

Exercise 6.1 *Gold fish can have translucent or pigmented scales. We will investigate how this property is inherited. One possibility is that it is inherited through one single gene A (monohybrids). If you cross gold fish with the gene combination Aa, you would then get the following distribution in the offspring:*

$$\begin{array}{ccc} AA & Aa & aa \\ 1/4 & 2/4 & 1/4 \end{array}$$

Since translucent scales are dominant, we would get 75% gold fish with translucent scales and 25% with pigmented scales. An experiment was made such that 100 gold fish were produced from parents with the Aa combination. The result was that 90 of the offspring had translucent scales. Test if these data are compatible with the "one-gene" hypothesis.

Exercise 6.2 *(Continuation of the previous exercise.) If the property is inherited through two genes (dihybridic), genetic theory would suggest the following distribution:*

$$\begin{array}{cccc} A\_B\_ & A\_bb & aaB\_ & aabb \\ 9/16 & 3/16 & 3/16 & 1/16 \end{array}$$

Only the combination aabb would give pigmented scales, i.e. we would expect 15/16 with translucent scales and 1/16 with pigmented scales. Use the data from the previous exercise to test whether the data are compatible with the hypothesis of dihybridic inheritance.

Exercise 6.3 *In a physiological experiment it was studied how a certain nutrient affected the duration of the life of rats. Among 225 animals of age one year, 144 were randomly selected to get the nutrient. The remaining 81 animals were used as controls. The number of animals that survived until the age of three years was counted. Results:*

	Exp. group	Control group	Total
Reached the age of 3	108	36	144
Did not reach 3 years	36	45	81
Total	144	81	225

Test (using a χ^2 test) if the nutrient has any effect on the life length.

Exercise 6.4 *An experiment on potatoes was made. The treatments were combinations of low and high amounts of nitrogen (a1 and a2), and*

6 Analysis of frequency data

low and high amounts of phosphorus (b1 and b2). Thus, there were four treatment combinations: a1b1, a1b2, a2b1 and a2b2. From each combination, 125 potatoes were analyzed. These were classified into one of four classes: A=no miscoloring; B=weak miscoloring; C=strongly miscolored; and D=very strongly miscolored. The experiment gave the following result:

	Treatment				
Miscoloring	a1b1	a1b2	a2b1	a2b2	Total
A	56	64	36	38	194
B	45	36	44	48	173
C	18	13	27	20	78
D	6	12	18	19	55
Total	125	125	125	125	500

Examine, using a χ^2 test, whether the treatments affect the degree of miscoloring of the potatoes.

Exercise 6.5 Is it dangerous to keep pet birds? Some data have been collected on 239 lung cancer patients, and on a control group of 429 healthy Berliners with the same age and sex distribution. One question was whether the person had a pet bird in the apartment. Results:

	Lung cancer patients	Healthy controls
Has a pet bird	98	101
No pet bird	141	328

Examine, using a χ^2 test, whether there is any relation between lung cancer and pet birds. State hypotheses, test statistics, and discuss the assumptions underlying the analysis. Is it dangerous to keep pet birds?

Exercise 6.6 In the general US population, the four blood groups O, A, B and AB are distributed as follows[1]:

O	A	B	AB
45%	40%	11%	4%

We will regard these as true percentages that are valid for the population. For a different ethnic group, a sample of 500 individuals gave

---

[1] Source: http://www.aabb.org/All_About_Blood/FAQs/aabb_faqs.htm

the following distribution of blood groups:

O	A	B	AB
228	206	40	26

A. Test the hypothesis that the blood group distribution for this ethnic group is the same as for the general US population.

The 500 individuals were also tested whether they were Rh positive or Rh negative. Results:

	O	A	B	AB
Rh+	198	176	28	22
Rh−	30	30	12	4

B. Test the hypothesis that the blood group distribution within this ethnic group is the same for Rh-positive as for Rh-negative individuals.

Exercise 6.7 Aird et al (1954) studied the relation between peptic ulcer and blood type. They obtained blood types for 1655 ulcer patients and for a sample of 10000 healthy controls. Results:

Blood type	Ulcer patients	Controls
O	911	4578
A	579	4219
B	124	890
AB	41	313
Total	1655	10000

Are there any relations between blood type and ulcer? Investigate this by making a χ^2 test. A complete solution should include hypotheses, test statistic, calculations, and a conclusion.

Chapter 7

Generalized linear models

7.1 Introduction

General linear models like t tests, ANOVA and regression analysis formally rest on the following set of assumptions:

- The residuals e_i are assumed to be normally distributed.

- The variance of the residuals is constant and equal to σ^2 for all treatments, i.e. the residuals are homoscedastic.

- The residuals are assumed to be independent.

In addition, the model used for the analysis is assumed to be correct.

Many formal and informal checks of these assumptions have been proposed. It has been suggested that ANOVA is rather robust against modest violations of the assumptions of normality and homoscedasticity. This means that the nominal level of the test, α, is not much affected if the distribution is slightly non-normal, or if the variances are slightly different, at least if the experiment is balanced.

However, there are situations where it is evident that the assumptions of normality and homoscedasticity do not hold. For such data, one approach is to base the analysis on some distribution other than the normal. A broad class of such analysis tools is called generalized linear models, which include distributions such as gamma, binomial and Poisson.

7.2 Generalized linear model theory

Generalized linear models (GLIMs) is a very general class of statistical models that includes many commonly used models as special cases. For example the class of general linear models (GLMs) that includes linear regression, analysis of variance and analysis of covariance, is a special case of GLIMs. GLIMs also include log-linear models for analysis of contingency tables, probit/logit regression, Poisson regression, and much more.

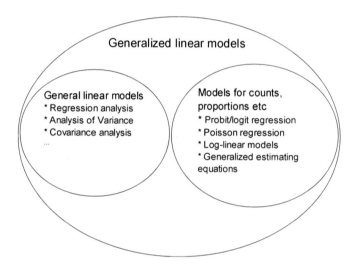

Generalized linear models provide a unified approach to modeling of many different types of response variables. In this section we will briefly review some aspects of generalized linear models. For a more complete coverage, reference is made to McCullagh and Nelder (1989) and Olsson (2002).

A general linear model may be written as

$$y_i = \beta_0 + \beta_1 x_{i1} + \beta_2 x_{i2} + \ldots + \beta_p x_{ip} + e_i$$

Let us denote

$$\eta_i = \beta_0 + \beta_1 x_{i1} + \beta_2 x_{i2} + \ldots + \beta_p x_{ip}$$

as the *linear predictor* part of the model. Generalized linear models are a generalization of general linear models in the following ways:

1. An assumption often made in a GLM is that the components y_i are

independently normally distributed with constant variance. We can relax this assumption to permit the distribution to be any distribution that belongs to the *exponential family* of distributions. This includes distributions such as normal, Poisson, gamma and binomial distributions.

2. Instead of modeling y (or its mean value μ) directly as a function of the linear predictor, we model some function $g(\mu)$ of μ. Thus, the model becomes

$$g(\mu_i) = \eta_i = \beta_0 + \beta_1 x_{i1} + \beta_2 x_{i2} + \ldots + \beta_p x_{ip}$$

The function $g(\cdot)$ is called a *link function*.
The specification of a generalized linear model thus involves:

1. specification of the distribution

2. specification of the link function $g(\cdot)$

3. specification of the linear predictor $\beta_0 + \beta_1 x_{i1} + \beta_2 x_{i2} + \ldots + \beta_p x_{ip}$

We will discuss these issues, starting with the distribution.

7.2.1 The exponential family of distributions

The exponential family is a general class of distributions that includes many well known distributions as special cases. It can be written in the form

$$f(y; \theta, \phi) = \exp\left[\frac{y\theta - b(\theta)}{a(\phi)} + c(y, \phi)\right]$$

where $a(\cdot)$, $b(\cdot)$ and $c(\cdot)$ are some functions. The so-called *canonical parameter* θ is some function of the location parameter of the distribution. Some authors differ between exponential family, where it is assumed that $a(\phi)$ is unity, and exponential dispersion family, which include the function $a(\phi)$ while assuming that the so called dispersion parameter ϕ is a constant; see Jørgensen (1987); Lindsey (1997, p. 10f). Certain link functions are, in a sense, "natural" for specific distributions. These are called the *canonical link*.

Some well-known distributions are, in fact, special cases of the exponential family. Some examples, along with their canonical links, are given in Table 7.1.

The parameters of generalized linear models are often estimated using the maximum likelihood method. Asymptotic standard errors of the estimators are obtained, along with a deviance statistic that can be used

7 Generalized linear models

Table 7.1: Some distributions in the Exponential family, along with their canonical link functions.

Distribution	Link	Type of data
Normal	Identity	Continuous, Normal
Binomial	Logit	Proportion
Poisson	Log	Count
Gamma	Inverse: $-\frac{1}{y}$	Lifetime
Exponential	Inverse	Lifetime
...		

to assess the fit of the model. The linear predictor has the same form as in general linear models.

The deviance statistic can be used to compare nested models, i.e. models where one model is a subset of a larger model. The difference in deviance between the two models can, in large samples, be used as a χ^2 statistic to test whether the extra parameters in the larger model are all equal to zero. The d.f. of the statistic is equal to the difference in d.f. between the two deviances.

Computer software for the analysis of generalized linear models are included e.g. in the S-plus, R and SAS packages. In the examples below, SAS was used.

7.3 Analysis of binary data

A binary response variable can take only two values, 0 and 1. Some examples of binary response variables are alive/dead, has a disease/does not have the disease, germinates/does not germinate, etc. Binary data may be presented in two ways. If several individuals have the same value on all covariates, the data may be grouped. In this case, the results can be presented as a proportion: "10 out of 40 patients survived". Alternatively, the data can be given separately for each individual with the value of the response recorded as 0 or 1. Such data can often be modeled using the binomial distribution. The object of modeling binary data is to find how the probability that $y = 1$ is related to one or more covariates. We will first study the case where there is only one continuous covariate, corresponding to simple linear regression. In the next section we will see an example corresponding to multiple regression.

Analysis of binary data

Table 7.2: Relation between the proportion of affected insects and the dose of rotenone. Data from Finney (1947).

Conc	Log(Conc)	No. of insects	No. affected	% affected
10.2	1.01	50	44	88
7.7	0.89	49	42	86
5.1	0.71	46	24	52
3.8	0.58	48	16	33
2.6	0.41	50	6	12

The probability of obtaining x "successes" among n independent binary (0 or 1) trials is

$$f(x) = \binom{n}{x} p^x (1-p)^{n-x}$$

for $x = 0, 1, 2, ..., n$. This is the probability function of the binomial distribution. The mean value and variance in the binomial distribution are

$$E(x) = np$$

and

$$Var(x) = np(1-p)$$

respectively.

7.3.1 Logistic regression

Example 7.1 *Finney (1947) reported on an experiment on the effect of rotenone, in different concentrations, when sprayed on the insect Macrosiphoniella sanborni, in batches of about fifty. For each individual insect it was recorded whether it was "affected" or not, so for each batch the proportion of insects affected was recorded. The results are given in Table 7.2.*

A plot of the relation between the proportion of affected insects and Log(Conc) is given in Figure 7.1. A fitted distribution is also included.

The data in Example 7.1 are an example of a "logistic regression" setting with grouped data. The dependent variable is a proportion. If we can assume independence within and between the different batches, the response can be assumed to follow a binomial distribution. According to Table 7.1, the canonical link for the binomial distribution is the logit

7 Generalized linear models

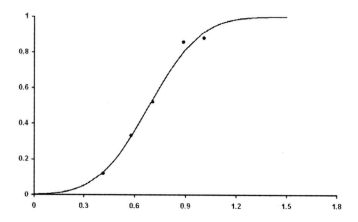

Figure 7.1: Relation between log(dose) and proportion of affected insect for Finney's data.

link, which can be written

$$g(p) = \log\left(\frac{p}{1-p}\right)$$

where log denotes the natural logarithm. This suggests that we may use a generalized linear model with a binomial distribution for the response and using a logit link. The model can then be written as

$$\log\left(\frac{p}{1-p}\right) = \beta_0 + \beta_1 x,$$

where x is log(conc); in dose-response studies, a log transformation of the dose often gives a better fit.

The following SAS program was used to analyze these data using Proc Genmod:

Analysis of binary data

```
DATA logit;
INPUT conc n x;
logconc=log10(conc);
CARDS;
10.2 50 44
 7.7 49 42
 5.1 46 24
 3.8 48 16
 2.6 50  6
;
PROC GENMOD data=logit;
  MODEL x/n = logconc /
  DIST=binomial
  LINK=logit;
RUN;
```

Part of the output is as follows:

```
                    The GENMOD Procedure

                      Model Information

        Data Set                          WORK.LOGIT
        Distribution                        Binomial
        Link Function                          Logit
        Response Variable (Events)                 x
        Response Variable (Trials)                 n


        Number of Observations Read            5
        Number of Observations Used            5
        Number of Events                     132
        Number of Trials                     243
```

This part simply gives us confirmation that we are using a binomial distribution and a logit link.

```
              Criteria For Assessing Goodness Of Fit

Criterion              DF            Value         Value/DF

Deviance                3           1.4241           0.4747
Scaled Deviance         3           1.4241           0.4747
Pearson Chi-Square      3           1.4218           0.4739
Scaled Pearson X2       3           1.4218           0.4739
Log Likelihood                   -119.8942
```

This section gives information about the fit of the model to the data. The scaled deviance indicates the fit of the model. Here, it can be interpreted as a χ^2 variate on 3 degrees of freedom, if the sample is large. In this case, the value is 1.42 which is clearly non-significant, indicating a good fit. Collett (1991) states that "a useful rule of thumb is that when the deviance on fitting a linear logistic model is approximately

151

7 Generalized linear models

equal to its degrees of freedom, the model is satisfactory" (p. 66).Thus, deviance/d.f. should not be much larger than 1.

Analysis Of Parameter Estimates

Parameter	DF	Estimate	Standard Error	Wald 95% Confidence Limits		Chi-Square	Pr > ChiSq
Intercept	1	-4.8869	0.6429	-6.1470	-3.6268	57.78	<.0001
logconc	1	7.1462	0.8928	5.3964	8.8959	64.07	<.0001
Scale	0	1.0000	0.0000	1.0000	1.0000		

The output finally contains an "analysis of parameter estimates". This gives estimates of model parameters, their standard errors, and a Wald test of each parameter in the form of a χ^2 test. Both the intercept and the slope are clearly significant ($p < 0.0001$). The estimated model is

$$\text{logit}(\widehat{p}) = \log\frac{\widehat{p}}{1-\widehat{p}} = -4.8869 + 7.1462 \cdot \log(\text{conc}).$$

The dose that affects 50% of the animals (ED_{50}) can easily be calculated: if $p = 0.5$ then $\log\frac{p}{1-p} = \log\frac{0.5}{0.5} = 0$ from which

$$\log(\text{conc}) = -\frac{\widehat{\beta}_0}{\widehat{\beta}_1} = \frac{4.8869}{7.1462} = 0.6839 \text{ giving}$$

$$\text{conc} = 10^{0.6839} = 4.83.$$

Instead of using the logit link we could have used the so called Probit link, i.e. the inverse normal distribution, for these data. The results are similar. Models using the probit link are called probit models; this was the analysis of the original publication (Finney, 1947).

7.3.2 Odds ratios

Suppose that two treatments against a disease (for example a common cold) are compared. The result for each patient might be that the disease is cured after one week ($y = 1$) or not cured ($y = 0$). A sample of patients is given each of the treatments. The parameters of interest in this example is the probability that the patient is cured, for each of the treatments (say p_1 and p_2).

An alternative to describing the data as probabilities is to use odds. The odds are defined as

$$\text{Odds} = \frac{p}{1-p}$$

for each treatment. Suppose, for example, that 100 patients get treat-

Analysis of binary data

ment 1 and 20 are cured within a week. The probability of cure can be estimated as $\hat{p} = \frac{20}{100} = 0.20$ and the odds of being cured is estimated as

$$\widehat{Odds}_1 = \frac{0.20}{0.80} = 0.250$$

Suppose that with the second treatment, 10 out of 100 patients are cured ($\hat{p}_2 = 0.1$). The odds for this group is estimated as

$$\widehat{Odds}_2 = \frac{0.1}{0.9} \approx 0.111$$

One way of comparing the treatments is to calculate the ratio between the odds, the so-called *odds ratio*:

$$\widehat{OR} = \frac{\widehat{Odds}_1}{\widehat{Odds}_2} = \frac{0.250}{0.111} = 2.252$$

The conclusion would be that it is more than twice as likely to get cured with treatment 1 than with treatment 2.

The parameter estimates in logistic regression models can be interpreted in terms of odds ratios. Consider the model

$$\log \frac{p}{1-p} = \beta_0 + \beta_1 x$$

If $x = 0$, we get $\log \frac{p_0}{1-p_0} = \beta_0$, i.e. the odds are $\frac{p_0}{1-p_0} = e^{\beta_0}$. If instead $x = 1$ we get $\log \frac{p_1}{1-p_1} = \beta_0 + \beta_1$, i.e. the odds are $\frac{p_1}{1-p_1} = e^{\beta_0+\beta_1}$. The odds ratio when comparing a group with $x = 1$, with a group with $x = 0$ is then

$$OR = \frac{\frac{p_1}{1-p_1}}{\frac{p_0}{1-p_0}} = \frac{e^{\beta_0+\beta_1}}{e^{\beta_0}} = e^{\beta_1}.$$

Thus, the value of the regression parameter β_1 can be used to estimate the odds ratio. This can be used when x is quantitative to compare the odds for different values of x. It is even more useful when x is a dummy variable.

7.3.3 Multiple logistic regression

In multiple logistic regression the response is a proportion or a binary variable. Such models may be analyzed by using a binomial distribution and a logistic link function. When there are many independent (x) variables, model building may be of primary interest.

Model building in multiple logistic regression models can be done in

7 Generalized linear models

essentially the same way as in standard multiple regression.

Example 7.2 *The data in Table 7.3, taken from Collett (1991), were collected to explore whether it was possible to diagnose nodal involvement in prostate cancer based on non-invasive methods. The variables are:*

Age	Age of the patient
Acid	Level of serum acid phosphate
X-ray	Result of x-ray examination (0=negative, 1=positive)
Size	Tumor size (0=small, 1=large)
Grade	Tumor grade (0=less serious, 1=more serious)
Inv	Nodal involvement (0=no, 1=yes)

The data analytic task is to explore whether the independent variables can be used to predict the probability of nodal involvement. We have two continuous covariates and three covariates coded as dummy variables. Initial analysis of the data suggests that the value of acid should be log-transformed prior to the analysis.

A useful rule-of-thumb in model building is to keep in the model all terms that are significant at, say, the 20% level. Alternatively, model building may be based on criteria like Akaike's information criterion; see Page 202. In this case, a kind of backward elimination process would start with the full model. We would then delete grade from the model ($p = 0.29$). In the model with age, log(acid), x-ray and size, age is not significant ($p = 0.26$). This suggests a model that includes log(acid), x-ray and size; in this model, all terms are significant ($p < 0.05$).

There are no indications of non-linear relations between log(acid) and the probability of nodal involvement. It remains to investigate whether any interactions between the terms in the model would improve the fit. To check this, interaction terms were added to the full model. Since there are five variables, the model was tested with all 10 possible pairwise interactions. The interactions size*grade ($p = 0.01$) and logacid*grade ($p = 0.10$) were judged to be strong enough for further consideration. Note that grade was not suggested by the analysis until the interactions were included. We then tried a model with both these interactions. Age could be deleted. The resulting model includes logacid ($p = 0.06$), x-ray ($p = 0.03$), size ($p = 0.21$), grade ($p = 0.19$), logacid*grade ($p = 0.11$), and size*grade ($p = 0.02$). A SAS program for the final model is as follows:

Table 7.3: Predictors of nodal involvement on prostate cancer patients

Age	Acid	Xray	Size	Grade	Inv
66	0.48	0	0	0	0
68	0.56	0	0	0	0
66	0.5	0	0	0	0
56	0.52	0	0	0	0
58	0.5	0	0	0	0
60	0.49	0	0	0	0
65	0.46	1	0	0	0
60	0.62	1	0	0	0
50	0.56	0	0	1	1
49	0.55	1	0	0	0
61	0.62	0	0	0	0
58	0.71	0	0	0	0
51	0.65	0	0	0	0
67	0.67	1	0	1	1
67	0.47	0	0	1	0
51	0.49	0	0	0	0
56	0.5	0	0	1	0
60	0.78	0	0	0	0
52	0.83	0	0	0	0
56	0.98	0	0	0	0
67	0.52	0	0	0	0
63	0.75	0	0	0	0
59	0.99	0	0	1	1
64	1.87	0	0	0	0
61	1.36	1	0	0	1
56	0.82	0	0	0	1
64	0.4	0	1	1	0
61	0.5	0	1	0	0
64	0.5	0	1	1	0
63	0.4	0	1	0	0
52	0.55	0	1	1	0
66	0.59	0	1	1	0
58	0.48	1	1	0	1
57	0.51	1	1	1	1
65	0.49	0	1	0	1
65	0.48	0	1	1	0
59	0.63	1	1	1	0
61	1.02	0	1	0	0
53	0.76	0	1	0	0
67	0.95	0	1	0	0
53	0.66	0	1	1	0
65	0.84	1	1	1	1
50	0.81	1	1	1	1
60	0.76	1	1	1	1
45	0.7	0	1	1	1
56	0.78	1	1	1	1
46	0.7	0	1	0	1
67	0.67	0	1	0	1
63	0.82	0	1	0	1
57	0.67	0	1	1	1
51	0.72	1	1	0	1
64	0.89	1	1	0	1
68	1.26	1	1	1	1

7 Generalized linear models

```
PROC GENMOD data=nodal;
CLASS xray size grade;
MODEL Inv = logacid xray size grade
            logacid*grade size*grade
            /DIST=Binomial;
RUN;
```
Part of the output is:

Criteria For Assessing Goodness Of Fit

Criterion	DF	Value	Value/DF
Deviance	46	36.2871	0.7889
Scaled Deviance	46	36.2871	0.7889
Pearson Chi-Square	46	42.7826	0.9301
Scaled Pearson X2	46	42.7826	0.9301
Log Likelihood	.	-18.1436	.

Analysis Of Parameter Estimates

Parameter			DF	Estimate	Std Err	ChiSquare	Pr>Chi
INTERCEPT			1	7.2391	3.4133	4.4980	0.0339
LOGACID			1	12.1345	6.5154	3.4686	0.0625
XRAY	0		1	-2.3404	1.0845	4.6571	0.0309
XRAY	1		0	0.0000	0.0000	.	.
SIZE	0		1	2.5098	2.0218	1.5410	0.2145
SIZE	1		0	0.0000	0.0000	.	.
GRADE	0		1	-4.3134	3.2696	1.7404	0.1871
GRADE	1		0	0.0000	0.0000	.	.
LOGACID*GRADE	0		1	-10.4260	6.6403	2.4652	0.1164
LOGACID*GRADE	1		0	0.0000	0.0000	.	.
SIZE*GRADE	0	0	1	-5.6477	2.4346	5.3814	0.0204
SIZE*GRADE	0	1	0	0.0000	0.0000	.	.
SIZE*GRADE	1	0	0	0.0000	0.0000	.	.
SIZE*GRADE	1	1	0	0.0000	0.0000	.	.
SCALE			0	1.0000	0.0000	.	.

The model fits well, with deviance/df =0.79. However, this should be interpreted with care: in cases where all individuals have different values on the covariates (i.e. when the data cannot be grouped), deviance/d.f. is not useful as a measure of model fit (Collett, 1991). Since the size*grade interaction is included in the model, the main effects of size and of grade should also be included. The output suggests the following models for grade 0 and 1, respectively:

Grade 0: $logit(\widehat{p}) = 2.93 + 1.71 \cdot \log(acid) - 2.34 \cdot x\text{-}ray - 3.14 \cdot size$
Grade 1: $logit(\widehat{p}) = 7.24 + 12.13 \cdot \log(acid) - 2.34 \cdot x\text{-}ray + 2.51 \cdot size$

The probability of nodal involvement increases with increasing acid level. The increase is higher for patients with serious (grade 1) tumors.

7.4 Analysis of counts: log-linear models

7.4.1 Two-way contingency tables

Count data can be summarized in the form of frequency tables, also called contingency tables. The data are then given as the number of observations with each combination of values of some categorical variables. Analysis of two-dimensional data of this kind often use so called χ^2 tests (see Chapter 6). We will now consider a more general type of analysis of such data, that is useful even for multi-dimensional tables. However, we will start by looking at a simple example with a contingency table of dimension 2×2.

Example 7.3 *Norton and Dunn (1985) studied possible relations between snoring and heart problems. For 2484 persons it was recorded whether the person had any heart problems and whether the person was a snorer. An interesting question is then whether there is any relation between snoring and heart problems. The data are as follows:*

	Snores		
Heart problems	Seldom	Often	Total
Yes	59	51	110
No	1958	416	2374
	2017	467	2484

We assume that the persons in the sample constitute a random sample from some population. Denote with p_{ij} the probability that a randomly selected person belongs to row category i and column category j of the table. This can be summarized as follows:

	Snores		
Heart problems	Seldom	Often	Total
Yes	p_{11}	p_{12}	$p_{1\cdot}$
No	p_{21}	p_{22}	$p_{2\cdot}$
	$p_{\cdot 1}$	$p_{\cdot 2}$	1

A dot in the subscript indicates a marginal probability. For example, $p_{\cdot 1}$ denotes the probability that a person snores seldom, i.e. $p_{\cdot 1} = p_{11} + p_{21}$.

7.4.2 A log-linear model for independence

If snoring and heart problems were statistically independent, it would hold that $p_{ij} = p_{i\cdot} p_{\cdot j}$ for all i and j. This is a model that we would

7 Generalized linear models

like to compare with the more general model that snoring and heart problems are dependent. Instead of modeling the probabilities, we can state the models in terms of expected frequencies $\mu_{ij} = np_{ij}$, where n is the total sample size and μ_{ij} is the expected number in cell (i,j). Thus, the independence model states that

$$\mu_{ij} = np_{i.}p_{.j}$$

This is a multiplicative model. By taking the logs of both sides we get an additive model assuming independence:

$$\begin{aligned}\log(\mu_{ij}) &= \log(n) + \log(p_{i.}) + \log(p_{.j}) \\ &= \mu + \alpha_i + \beta_j.\end{aligned}$$

Here, α_i denotes the row effect (i.e. the effect of variable A), and β_j denotes the column effect (i.e. the effect of variable B). In log-linear model literature, effects are often denoted with symbols like λ_i^X, but we keep a notation that is in line with the notation of previous chapters. We can see that this model is a linear model (a linear predictor), and that the link function is log. Models of this type are called log-linear models.

Note that a model for a cross-table of dimension $r \times c$ can include at most $(r-1)$ parameters for the row effects and $(c-1)$ parameters for the column effects. This is analogous to ANOVA models. One way to constrain the parameters is to set the last parameter of each kind equal to zero. In our example, $r = c = 2$ so we need only one parameter α_i and one β_j, for example α_1 and β_1. In GLIM terms, the model for our example data can then be written as

$$\begin{aligned}\log(\mu_{11}) &= \mu + 1 \cdot \alpha_1 + 1 \cdot \beta_1 \\ \log(\mu_{12}) &= \mu + 1 \cdot \alpha_1 + 0 \cdot \beta_1 \\ \log(\mu_{21}) &= \mu + 0 \cdot \alpha_1 + 1 \cdot \beta_1 \\ \log(\mu_{22}) &= \mu + 0 \cdot \alpha_1 + 0 \cdot \beta_1\end{aligned}$$

7.4.3 When independence does not hold

If independence does not hold we need to include in the model terms of type $(\alpha\beta)_{ij}$ that account for the dependence. The terms $(\alpha\beta)_{ij}$ represent interaction between the factors A and B, i.e. the effect of one variable depends on the level of the other variable. Then the model becomes

$$\log(\mu_{ij}) = \mu + \alpha_i + \beta_j + (\alpha\beta)_{ij}.$$

Any two-dimensional cross table can be perfectly represented by this type of model; this model is called the *saturated model*. We can test the restrictions imposed by removing the parameters $(\alpha\beta)_{ij}$ by comparing the deviances: the saturated model will have deviance 0 on 0 degrees of freedom, so the deviance from fitting the restricted model can be used directly to test the hypothesis of independence.

7.4.4 Distributions for count data

So far, we have seen that a model for the expected frequencies in a crosstable can be formulated as a log-linear model. This model has the following properties:

The predictor is a linear predictor of the same type as in ANOVA.

The link function is a log function.

The distribution is a Poisson distribution.

This can be explained as follows. Suppose that a nominal variable y has k distinct values $y_1, y_2, ..., y_k$. We observe counts $n_1, n_2, ..., n_k$. The expected number of observations in cell i is μ_i. If the observations arrive randomly, the probability to observe n_i observations in cell i is

$$p(n_i) = \frac{e^{-\mu_i} \mu_i^{n_i}}{n_i!}$$

which is the probability function of a Poisson distribution. Note that in this case, the total sample size n is not regarded as fixed. Since sums of Poisson variables follow a Poisson distribution, n in itself follows a Poisson distribution with mean value $\sum_{i=1}^{k} \mu_i$.

The probability function of a Poisson distribution is

$$f(x) = \frac{\lambda^x e^{-\lambda}}{x!}$$

for $x = 0, 1, 2, ...$ The Poisson distribution has mean value and variance

$$E(x) = \lambda$$

and

$$Var(x) = \lambda.$$

7.4.5 Relation to contingency tables

Contingency tables can be of many different types. In some cases, the total sample size is fixed; an example is when it has been decided that $n = 1000$ individuals will be interviewed about some political question. In some cases even some of the margins of the table may be fixed. An example is when 500 males and 500 females will participate in a survey. A table with a fixed total sample size would suggest a multinomial distribution; if in addition one or more of the margins are fixed we would assume a product multinomial distribution. However, as noted by Agresti (1996), "For most analyses, one need not worry about which sampling model makes the most sense. For the primary inferential methods in this text, the same results occur for the Poisson, multinomial and independent binomial/multinomial sampling models" (p. 19).

Suppose that we observe a contingency table of size $i \times j$. The probability that an observation will fall into cell (i, j) is p_{ij}. If the observations are independent and arrive randomly, the number of observations falling into cell (i, j) follows a Poisson distribution with mean value μ_{ij}, if the total sample size n is random. If the cell counts n_{ij} follow a Poisson distribution then the conditional distribution of $n_{ij}|n$ is multinomial. The Poisson distribution is often used to model count data since it is rather easy to handle.

Note, however, that there is no guarantee that a given set of data will adhere to this assumption. Sometimes the data may show a tendency to "cluster" so that the arrival of one observation in a specific cell may increase the probability that the next observation falls into the same cell. This would lead to overdispersion; a distribution called the negative binomial distribution may be used in some such cases; see section 7.9. For the moment, however, we will see what happens if we tentatively accept the Poisson assumption for the data on snoring and heart problems.

7.4.6 Analysis of the example data

We analyzed the data on page 157 using the Genmod procedure with Poisson distribution and a log link. The program was:

```
DATA snoring;
INPUT snore heart count;
CARDS;
1 1   51
1 0   416
0 1   59
0 0 1958
;
PROC GENMOD DATA=snoring;
CLASS snore heart;
MODEL count = snore heart /
         LINK = log
         DIST = poisson
;
RUN;
```

The output contains the following information:

```
           Criteria For Assessing Goodness Of Fit

   Criterion              DF      Value        Value/DF

   Deviance               1       45.7191      45.7191
   Scaled Deviance        1       45.7191      45.7191
   Pearson Chi-Square     1       57.2805      57.2805
   Scaled Pearson X2      1       57.2805      57.2805
   Log Likelihood         .    15284.0145         .
```

```
               Analysis Of Parameter Estimates

   Parameter        DF   Estimate   Std Err   ChiSquare   Pr>Chi

   INTERCEPT        1    3.0292     0.1041    847.2987    0.0001
   SNORE      0     1    1.4630     0.0514    811.6746    0.0001
   SNORE      1     0    0.0000     0.0000       .          .
   HEART      0     1    3.0719     0.0975    992.0240    0.0001
   HEART      1     0    0.0000     0.0000       .          .
   SCALE            0    1.0000     0.0000       .          .
```

NOTE: The scale parameter was held fixed.

A similar analysis that includes an interaction term would produce a deviance of 0 on 0 $d.f.$. Thus, the difference between our model and the saturated model can be tested; the difference in deviance is 45.7 on 1 degree of freedom which is highly significant when compared with the χ^2 limit with 1 $d.f$; the 5% limit is is 3.84 and the 0.1% limit is 10.828. We conclude that snoring and heart problems do not seem to

7 Generalized linear models

be independent. Note that the Pearson chi-square of 57.28 on 1 d.f. presented in the output is based on the textbook formula

$$\chi^2 = \sum_{i,j} \frac{\left(n_{ij} - \widehat{\mu}_{ij}\right)^2}{\widehat{\mu}_{ij}}.$$

The conclusion is the same, in this case, but the deviance and the Pearson tests are not quite identical.

The output also gives us estimates of the three parameters of the model: $\widehat{\mu} = 3.0292$, $\widehat{\alpha}_1 = 1.4630$ and $\widehat{\beta}_1 = 3.0719$. An analysis of the saturated model would give an estimate of the interaction parameter as $\widehat{(\alpha\beta)}_{11} = 1.4033$. From this we can calculate the odds ratio OR as

$$OR = \exp(1.4033) = 4.07.$$

Patients who snore have four times larger odds of having heart problems. Odds ratios in log-linear models are further discussed in Section 7.4.10.

7.4.7 Higher-order tables

The arguments used above for the analysis of two-dimensional contingency tables can be generalized to tables of higher order. A general (saturated) model for a three-way table can be written as

$$\log\left(\mu_{ijk}\right) = \mu + \alpha_i + \beta_j + \gamma_k + (\alpha\beta)_{ij} + (\alpha\gamma)_{ik} + (\beta\gamma)_{jk} + (\alpha\beta\gamma)_{ijk}$$

An important part of the analysis is to decide which terms to include in the model.

Example 7.4 *Table 7.4 contains data from a survey at Wright State University in 1992[1]. 2276 high school seniors were asked whether they had ever used alcohol (A), cigarettes (C) and/or marijuana (M). This is a three-way contingency table of dimension $2 \times 2 \times 2$.*

7.4.8 Types of independence

Models for data of the type given in Example 7.4 can include the main effects of A, C and M and different interactions containing these. The presence of an interaction, for example A*C, means that students who use alcohol have a higher (or lower) probability of also using cigarettes.

---

[1] Data quoted from Agresti (1996) who credited the data to Professor Harry Khamis.

Analysis of counts: log-linear models

Table 7.4: Use of alcohol, cigarettes and marijuana at Wright State University in 1992. Data from Agresti (1996).

Alcohol use	Cigarette use	Marijuana use Yes	No
Yes	Yes	911	538
	No	44	456
No	Yes	3	43
	No	2	279

One way of interpreting interactions is to calculate odds ratios; we will return to this topic soon.

A model of type A C M A*C A*M would permit interaction between A and C, and between A and M, but not between C and M. C and M are then said to be conditionally independent, controlling for A.

A model that only contains the main effects, i.e. the model A C M is called a mutual independence model. In this example this would mean that use of one drug does not change the risk of using any other drug.

A model that contains all interactions up to a certain level, but no higher-order interactions, is called a homogeneous association model.

7.4.9 Genmod analysis of the drug use data

The saturated model that contains all main effects and all two- and three way interactions was fitted to the data as a baseline. The three-way interaction A*C*M was not significant ($p = 0.53$). The output for the homogeneous association model containing all two-way interactions was as follows:

```
            Criteria For Assessing Goodness Of Fit

Criterion              DF         Value       Value/DF

Deviance                1        0.3740         0.3740
Scaled Deviance         1        0.3740         0.3740
Pearson Chi-Square      1        0.4011         0.4011
Scaled Pearson X2       1        0.4011         0.4011
Log Likelihood          .    12010.6124              .
```

The fit of this model is good; a simple rule of thumb is that Value/$d.f.$ should not be too much larger than 1. The parameter estimates for this model are as follows:

7 Generalized linear models

```
              Analysis Of Parameter Estimates

Parameter                DF    Estimate    Std Err     ChiSquare    Pr>Chi

INTERCEPT                 1     6.8139     0.0331    42312.0532     0.0001
A          No             1    -5.5283     0.4522      149.4518     0.0001
A          Yes            0     0.0000     0.0000            .           .
C          No             1    -3.0158     0.1516      395.6463     0.0001
C          Yes            0     0.0000     0.0000            .           .
M          No             1    -0.5249     0.0543       93.4854     0.0001
M          Yes            0     0.0000     0.0000            .           .
A*C        No   No        1     2.0545     0.1741      139.3180     0.0001
A*C        No   Yes       0     0.0000     0.0000            .           .
A*C        Yes  No        0     0.0000     0.0000            .           .
A*C        Yes  Yes       0     0.0000     0.0000            .           .
A*M        No   No        1     2.9860     0.4647       41.2933     0.0001
A*M        No   Yes       0     0.0000     0.0000            .           .
A*M        Yes  No        0     0.0000     0.0000            .           .
A*M        Yes  Yes       0     0.0000     0.0000            .           .
C*M        No   No        1     2.8479     0.1638      302.1409     0.0001
C*M        No   Yes       0     0.0000     0.0000            .           .
C*M        Yes  No        0     0.0000     0.0000            .           .
C*M        Yes  Yes       0     0.0000     0.0000            .           .
SCALE                     0     1.0000     0.0000            .           .
```

All remaining interactions in the model are highly significant which means that no further simplification of the model is suggested by the data.

7.4.10 Interpretation through odds ratios

Consider, for the moment, a $2 \times 2 \times k$ cross table of variables x, y and z. Within a fixed level j of z, the conditional odds ratio for describing the relationship between x and y is

$$\theta_{xy(j)} = \frac{\mu_{11j}\mu_{22j}}{\mu_{12j}\mu_{21j}}$$

where μ denotes expected values. In contrast, in the marginal odds ratio the value of the variable z is ignored and we calculate the odds ratio as

$$\theta_{xy} = \frac{\mu_{11\cdot}\mu_{22\cdot}}{\mu_{12\cdot}\mu_{21\cdot}} \qquad (7.1)$$

where the dot indicates summation over all levels of z. The odds ratios can be estimated from the parameter estimates; it holds that, for example,

$$\widehat{\theta}_{xy} = \exp\left[\widehat{(\alpha\beta)}_{11} + \widehat{(\alpha\beta)}_{22} - \widehat{(\alpha\beta)}_{12} - \widehat{(\alpha\beta)}_{21}\right]$$

In our drug use example, the chosen model does not contain any three-way interaction, and only one parameter is estimable for each interaction. Thus, the partial odds ratios for the two-way interactions can be estimated as:

A*C: $\exp(2.0545) = 7.80$

A*M: $\exp(2.9860) = 19.81$

C*M: $\exp(2.8479) = 17.25$

As an example of an interpretation, a student who has tried alcohol has an odds of also having tried marijuana of 19.81, regardless of reported cigarette use.

7.5 Poisson regression

Responses in the form of counts can sometimes be modeled as a function of a numerical covariate. The resulting models are called Poisson regression models.

Example 7.5 *Table 7.5, taken from Haberman (1978), shows the distribution of stressful events reported by 147 subjects who have experienced exactly one stressful event. The table gives the number of persons reporting a stressful event 1, 2, ..., 18 months prior to the interview. We want to model the occurrence of stressful events as a function of time.*

Table 7.5: Distribution of stressful events experienced by 147 subjects. Data from Haberman (1978).

Months	1	2	3	4	5	6	7	8	9
Number	15	11	14	17	5	11	10	4	8
Months	10	11	12	13	14	15	16	17	18
Number	10	7	9	11	3	6	1	1	4

One approach to modeling the occurrence of stressful events as a function of $x=$months is to assume that the number of persons responding for any month is a Poisson variate. The canonical link for the Poisson distribution is log, so a first attempt at modelling these data is to assume that

$$\log(\mu) = \beta_0 + \beta_1 x \qquad (7.2)$$

7 Generalized linear models

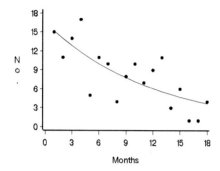

Figure 7.2: Distribution of persons remembering stressful events

This is a generalized linear model with a Poisson distribution, a log link and a simple linear predictor. A SAS program for this model can be written as

```
DATA stress;
INPUT months number @@;
CARDS;
 1 15   2 11   3 14   4 17   5  5   6 11   7 10   8  4   9  8  10 10
11  7  12  9  13 11  14  3  15   6 16   1 17   1 18   4
;
PROC GENMOD DATA=stress;
MODEL number = months / DIST=poisson LINK=log OBSTATS RESIDUALS;
MAKE 'obstats' out=ut;
RUN;
```

Part of the output is

```
              Criteria For Assessing Goodness Of Fit

              Criterion              DF           Value      Value/DF

              Deviance               16         24.5704        1.5356
              Scaled Deviance        16         24.5704        1.5356
              Pearson Chi-Square     16         22.7145        1.4197
              Scaled Pearson X2      16         22.7145        1.4197
              Log Likelihood          .        174.8451             .
```

The data have a less than perfect fit to the model, with Value/$d.f.$=1.53; the p value is 0.078. We find that the memory of stressful events fades away as $\log(\mu) = 2.80 - 0.084x$. A plot of the data, along with the fitted regression line, is given as Figure 7.2. Figure 7.3 shows the data and regression line with a log scale for the y axis.

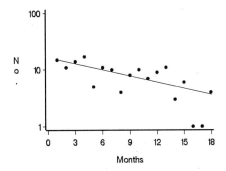

Figure 7.3: Distribution of persons remembering stressful events; log scale

7.6 Rate data

Events that may be assumed to be essentially Poisson distributed are sometimes recorded on units of different size. For example, the number of crimes recorded in a number of cities may depend on the size of the city, such that "number of crimes per 1000 inhabitants" is a meaningful measure of crime rate. Data of this type are called *rate data*.

If we denote the measure of size of each unit with t, we can model this type of data as

$$\log\left(\frac{\mu}{t}\right) = \beta_0 + \beta_1 x_1 + \beta_2 x_2 \ldots \beta_p x_p$$

i.e. as a linear predictor with p covariates $x_1, x_2, \ldots, x_p$. This means that

$$\log(\mu) = \log(t) + \beta_0 + \beta_1 x_1 + \beta_2 x_2 \ldots \beta_p x_p \qquad (7.3)$$

The adjustment term $\log(t)$ is called an *offset*. This is a known constant for each observation. The offset can easily be included in models analyzed with e.g. Proc Genmod.

Example 7.6 *Doll and Hill (1954) sent a questionnaire to all British doctors in 1951. One of the questions was whether they smoked tobacco. Later, data on the deaths of the doctors was collected. The data are given in Table 7.6.*

The data can be modeled using a Poisson distribution, using the logarithm of the number of person years as offset. A SAS program can be

7 Generalized linear models

Table 7.6: Smoking habits and deaths of British medical doctors. From Dole and Hill (1954)

Age group	Smokers Deaths	Smokers Person years	Non-smokers Deaths	Non-smokers Person years
35-44	32	52407	2	18790
45-54	104	43248	12	10673
55-64	206	28612	28	5710
65-74	186	12663	28	2585
75-84	102	5317	31	1462

written as follows:
```
DATA Doll;
INPUT age deaths personyears smok;
lpy=log(personyears);
cards;
40    32     52407  1
50    104    43248  1
60    206    28612  1
70    186    12663  1
80    102    5317   1
40    2      18790  0
50    12     10673  0
60    28     5710   0
70    28     2585   0
80    31     1462   0
;
PROC GENMOD data=doll;
CLASS age smok;
MODEL deaths = age smok age*smok/
     dist=poisson offset=lpy type3;
RUN;
```

Some of the output is as follows:

LR Statistics For Type 3 Analysis

Source	DF	Chi-Square	Pr > ChiSq
age	4	536.69	<.0001
smok	1	18.95	<.0001
age*smok	4	12.13	0.0164

The results indicate that age (as expected) and smoking habits are related to the risk of dying. Some interaction between age and smoking is also present. The relation can be illustrated as in Figure 7.4. The death rate is higher for smokers in all age groups exept the last.

Ordinal data

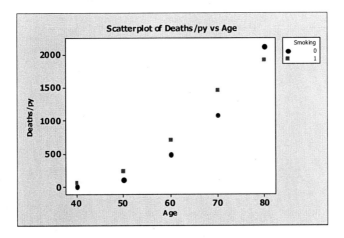

Figure 7.4: Relation between death risk and age for smokers and for non-smokers. Data from Doll and Hill (1954).

7.7 Ordinal data

Ordinal data are data where the response variable is categorical with ordered categories. Examples of such data are school marks, answers to attitude items (Agrees, Undecided, Disagrees), and doctor's judgements of disease severity.

Ordinal data can be analyzed using generalized linear models using two seemingly different arguments: latent variables, or proportional odds.

Latent variables One way to model the response variable is to assume that the observed ordinal variable y is related to some underlying, latent, variable η through a relation of type

$$\begin{aligned} y &= 1 &&\text{if} & \eta &< \tau_1 \\ y &= 2 &&\text{if} & \tau_1 &\leq \eta < \tau_2 \\ &\vdots \\ y &= s &&\text{if} & \tau_{s-1} &\leq \eta \end{aligned}$$

An example of this point of view is illustrated in Figure 7.5, where the latent variable is assumed to have a symmetric distribution, for example a logistic or a normal distribution.

7 Generalized linear models

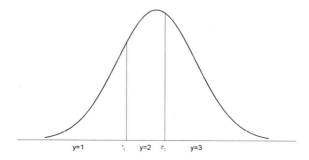

Figure 7.5: Ordinal variable with three scale steps generated by cutting a continuous variable at two thresholds

Proportional odds The proportional odds model for an ordinal response variable is a model for cumulative probabilities of type

$$P(y \leq j) = p_1 + p_2 + \ldots + p_j$$

where for simplicity we index the categories of the response variable with integers. The cumulative logits are defined as

$$\text{logit}(P(y \leq j)) = \log \frac{P(y \leq j)}{1 - P(y \leq j)}$$

The cumulative logits are defined for each of the categories of the response except the first one. Thus, for a response variable with 5 categories we would get $5 - 1 = 4$ different cumulative logits.

The proportional odds model for ordinal response suggests that all these cumulative logit functions can be modeled as

$$\text{logit}(P(y \leq j)) = \alpha_j + \beta x$$

i.e. the functions have different intercepts α_i but a common slope β. This means that the odds ratio, for two different values x_1 and x_2 of the predictor x, has the form

$$\frac{P(y \leq j | x_2) / P(y > j | x_2)}{P(y \leq j | x_1) / P(y > j | x_1)}.$$

The log of this odds ratio equals $\beta(x_2 - x_1)$, i.e. the log odds is pro-

Table 7.7: Ordinal data on the response to treatment for arthritis pain. From Koch and Edwards (1988).

Gender	Treatment	Response		
		Marked	Some	None
Female	Active	16	5	6
Female	Placebo	6	7	19
Male	Active	5	2	7
Male	Placebo	1	0	10

portional to the difference between x_2 and x_1. This is why the model is called the proportional odds model. Instead of the cumulative logit, a cumulative normal or a cumulative Gumbel distribution can be used.

It can be shown that the "latent variable" approach leads to the same model as the proportional odds model (Agresti, 1984). In the Genmod or Logistic procedures in SAS it is possible to specify the form of the link function to be logistic, complementary log-log, or normal. This leads to a class of models called ordinal regression models, for example ordinal logit regression or ordinal probit regression. Ordinal regression models can also be handled by Minitab.

Example 7.7 *Koch and Edwards (1988) considered analysis of data from a clinical trial on the response to treatment against arthritis pain. The data are given in Table 7.7.*

The object is to model the response as a function of gender and treatment. We will attempt a proportional odds model for the cumulative logits, using the Genmod procedure of SAS (2008).

The data were input in a form where the data lines had the form
F A 3 16
F A 2 5
...
The program was written as follows:

```
PROC GENMOD data=Koch order=formatted;
   CLASS gender treat;
   FREQ count;
   MODEL response = gender treat gender*treat/
       LINK=cumlogit aggregate=response TYPE3;
RUN;
```

7 *Generalized linear models*

Part of the output was:

```
              Analysis Of Parameter Estimates

                                   Standard      Wald 95%          Chi-
Parameter              DF  Estimate   Error  Confidence Limits    Square

Intercept1              1   3.6746   1.0125   1.6901   5.6591     13.17
Intercept2              1   4.5251   1.0341   2.4983   6.5519     19.15
gender       F          1  -3.2358   1.0710  -5.3350  -1.1366      9.13
gender       M          0   0.0000   0.0000   0.0000   0.0000        .
treat        A          1  -3.7826   1.1390  -6.0150  -1.5503     11.03
treat        P          0   0.0000   0.0000   0.0000   0.0000        .
gender*treat F  A       1   2.1110   1.2461  -0.3312   4.5533      2.87
gender*treat F  P       0   0.0000   0.0000   0.0000   0.0000        .
gender*treat M  A       0   0.0000   0.0000   0.0000   0.0000        .
gender*treat M  P       0   0.0000   0.0000   0.0000   0.0000        .
Scale                   0   1.0000   0.0000   1.0000   1.0000
```

```
         LR Statistics For Type 3 Analysis

                                Chi-
         Source          DF    Square    Pr > ChiSq

         gender           1     18.01      <.0001
         treat            1     28.15      <.0001
         gender*treat     1      3.60      0.0579
```

We can note that there is a slight (but not significant) interaction; that there are significant gender differences and that the treatment has a significant effect. The signs of the parameters indicate that patients on active treatment experienced a higher degree of pain relief and that the females experienced better pain relief than the males. The cumulative probit model gave similar results except that the interaction term was further from being significant ($p = 0.11$).

7.8 Gamma distribution

The Gamma distribution is often used to model duration data, i.e. data that measure how long it takes for a certain event to occur. Some examples are data on lifetimes, time until breakdown of a machine, etc.

Example 7.8 *Hurn et al (1945), quoted from McCullagh and Nelder (1989), studied the clotting time of blood. Two different clotting agents*

Gamma distribution

were compared for different concentrations of plasma. The data are:

	Clotting time	
Conc	Agent 1	Agent 2
5	118	69
10	58	35
15	42	26
20	35	21
30	27	18
40	25	16
60	21	13
80	19	12
100	18	12

Duration data can often be modeled using the gamma distribution. The canonical link of the gamma distribution is minus the inverse link, $-1/\mu$. Preliminary analysis of the data suggested that the relation between clotting time and concentration was better approximated by a linear function if the concentrations were log-transformed. Thus, the models that were fitted to the data were of type

$$\frac{1}{\mu} = \beta_0 + \beta_1 d + \beta_2 x + \beta_3 dx$$

where $x = \log(\text{conc})$ and d is a dummy variable with $d = 1$ for lot 1 and $d = 0$ for lot 2. This is a kind of covariance analysis model. A Genmod analysis of the full model gave the following output:

```
          Criteria For Assessing Goodness Of Fit

Criterion              DF         Value        Value/DF

Deviance               14        0.0294          0.0021
Scaled Deviance        14       17.9674          1.2834
Pearson Chi-Square     14        0.0298          0.0021
Scaled Pearson X2      14       18.2205          1.3015
Log Likelihood          .      -26.5976               .
```

7 *Generalized linear models*

```
              Analysis Of Parameter Estimates

Parameter        DF    Estimate    Std Err    ChiSquare   Pr>Chi

INTERCEPT         1     -0.0239     0.0013    359.9825    0.0001
AGENT       1     1      0.0074     0.0015     24.9927    0.0001
AGENT       2     0      0.0000     0.0000         .          .
LC                1      0.0236     0.0005   1855.0452    0.0001
LC*AGENT    1     1     -0.0083     0.0006    164.0704    0.0001
LC*AGENT    2     0      0.0000     0.0000         .          .
SCALE             1    611.1058   203.6464         .          .
```

NOTE: The scale parameter was estimated by maximum likelihood.

We can see that all parameters are significantly different from zero, which means that we cannot simplify the model any further. The scaled deviance is 17.97 on 14 *df*. A plot of the fitted model, along with the data, is given in Figure 7.6. The fit is good, but McCullagh and Nelder note that the lowest concentration value might have been misrecorded.

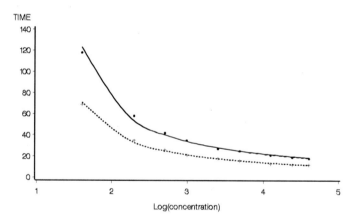

Figure 7.6: Relation between clotting time and log(concentration)

7.9 Over-dispersion

It may sometimes occur that a generalized linear model has a bad fit to the data. This is visible as a large value of the deviance, for example a deviance that is significant when interpreted as a Chi-square variate with d.f. taken from the model. A common rule of thumb that is valid for many (but not all) types of models is that Deviance/d.f. should not

Table 7.8: Somatic cell count in milk from sheep under two treatments

M										
1	2 966	269	59	1 887	3 452	189	93	618	130	2 493
2	186	107	65	126	123	164	408	324	548	139

be too much larger than 1, for a well-fitting model. A large value of Deviance/d.f. indicates over-dispersion. This means that the variation in the data is larger than expected, if the chosen model is correct. For example, for a Poisson distribution, the mean value is equal to the variance. If the variance in the data is much larger than the mean value, this indicates over-dispersion.

Over-dispersion indicates problems with the chosen model. The problem might be that the observations are clustered, i.e. that groups of observations are dependent. The effect of over-dispersion is that p values become too small: it is too easy to get significant results, even if no true differences exist.

We will not discuss over-dispersion in detail in this short review; for a more thorough discussion see for example McCullagh and Nelder (1989) or Olsson (2002). However, a few methods for handling over-dispersion can be briefly mentioned.

1. Introduce an overdispersion parameter into the model. This can be done by asking the program to use deviance/d.f. as an estimator of the amount of overdispersion and re-estimate the model, thus forcing dev/d.f. to be 1. This is a method that is completely heuristic, but that may be used if other alternatives fail.

2. Choose some other distribution for which the variance is larger than for the chosen distribution. For example, overdispersed Poisson data may sometimes be modeled using the negative binomial distribution, and overdispersed binomial data may be better fit using a beta-binomial distribution.

3. A third alternative is to use a method for estimating the variance that is robust to deviations from the distributional assumptions. Some computer programs include a so-called Sandwich estimator that has this property.

Example 7.9 *The data in Table 7.8 presents counts of somatic cells in the milk of sheep. The sheep were milked manually or mechanically. The question is whether the milking method affects the cell count.*

A Poisson model was attempted first. The Genmod program is:

7 Generalized linear models

```
PROC GENMOD data=sheep;
CLASS method;
MODEL cellcount = method
/ dist=poisson;
RUN;
```

The results (not shown) indicate that there is a strong overdispersion: (deviance/d.f.=825.7). The test of the treatment effect is highly significant ($p < 0.0001$).

The second model asks the program to use Deviance/d.f. as a scale parameter. This is done by changing the third line in the program to

```
/ dist=poisson scale=deviance;
```

Now, the fit is (of course) good, with deviance=d.f.=1. The type 3 test is

LR Statistics For Type 3 Analysis

Source	Num DF	Den DF	F Value	Pr > F	Chi-Square	Pr > ChiSq
method	1	18	9.24	0.0071	9.24	0.0024

Another option is to use a negative binomial distribution. This is obtained by replacing DIST=POISSON with DIST=NEGBIN in the program. Parts of the results are

Criteria For Assessing Goodness Of Fit

Criterion	DF	Value	Value/DF
Deviance	18	22.8801	1.2711
Scaled Deviance	18	22.8801	1.2711
Pearson Chi-Square	18	16.4703	0.9150
Scaled Pearson X2	18	16.4703	0.9150

LR Statistics For Type 3 Analysis

Source	DF	Chi-Square	Pr > ChiSq
method	1	11.18	0.0008

Finally, one analysis was made that produces robust "sandwich" estimators which are less sensitive to errors in the model specification. The program is as follows. obs is a variable that gives the observation number.

```
PROC GENMOD data=sheep;
CLASS method obs;
MODEL count = method /DIST=Poisson  type3 ;
REPEATED  subject=obs;
RUN;
```

Results:

```
              GEE Fit Criteria

                QIC      -206.5494
                QICu     -206.1494

         Score Statistics For Type 3 GEE Analysis

                              Chi-
         Source        DF    Square    Pr > ChiSq

         method         1     4.64       0.0313
```

The analyses produce different p values for the test of treatment effect. The Poisson analysis gives $p < 0.0001$ while the robust analysis gives $p = 0.0313$ and the negative binomial model gives $p = 0.0008$. The Poisson analysis has probably exaggerated the effect, while the robust analysis might be a bit conservative. If the assumptions underlying the negative binomial model are valid, it might be the best choice.

7 Generalized linear models

7.10 Exercises

Exercise 7.1 *Freeman (1989) studied the relation between the smoking habits of the mother and the survival of newborn babies. 4915 children were monitored during their first year, and it was recorded whether the child died (y=1) or not (y=0). The mothers were classified as smokers (x=1) or non-smokers (x=0). A logistic regression was made according to the following model:*

$$log(p/(1-p)) = a + bx$$

A computer program produced the following output:

```
            Estimate        Standard error
Intercept   -4.0686         0.1172
Smoking      0.5640         0.2871
```

A. Is there any significant relation between the smoking habits of the mother and the risk that the baby dies during its first year? Perform a test at the 5% level.

B. What is the direction of the relation? Use the printout to calculate the estimated probability of survival for children of smokers, and for children of non-smokers.

Exercise 7.2 *Even before the space shuttle Challenger exploded on January 20, 1986, NASA had collected data from 23 earlier launches. One part of these data was the number of O-rings that had been damaged at each launch. O-rings are a kind of gaskets that will prevent hot gas from leaking during takeoff. The data included the number of damaged O-rings, and the temperature (in Fahrenheit) at the time of the launch. On the fateful day when the Challenger exploded, the temperature was $31°F$.*

One might ask whether the probability that an O-ring is damaged is related to the temperature. The following data are available:

Exercises

No. of defective O-rings	Temperature °F	No. of defective O-rings	Temperature °F
2	53	0	70
1	57	1	70
1	58	1	70
1	63	0	72
0	66	0	73
0	67	0	75
0	67	2	75
0	67	0	76
0	68	0	76
0	69	0	78
0	70	0	79
		0	81

A statistician fitted a generalized linear model to these data. The model used a Poisson distribution and a log link: where μ is the mean value of the Poisson distribution and x is the temperature. Parts of the SAS/Genmod output from this analysis are presented below.

```
              Criteria For Assessing Goodness Of Fit

        Criterion              DF          Value       Value/DF

        Deviance               21         16.8337       0.8016
        Scaled Deviance        21         16.8337       0.8016
        Pearson Chi-Square     21         28.1745       1.3416
        Scaled Pearson X2      21         28.1745       1.3416
        Log Likelihood                   -14.6442

                 Analysis Of Parameter Estimates

                          Standard    Wald 95% Confidence    Chi-
Parameter   DF  Estimate    Error         Limits            Square   Pr > ChiSq

Intercept    1   5.9691    2.7628
Temp         1  -0.1034    0.0430
Scale        0   1.0000    0.0000
```

A. Test whether temperature has any significant effect on the failure of O-rings.

B. Predict the outcome of the response variable if the temperature is $31°F$.

C. Estimate the probability that three or more O-rings fail if the temperature is $31°F$.

Exercise 7.3 *Potatoes were grown using low or high nitrogen (A1 and A2) and low or high level of phosphorus (B1 or B2). The response was degree of miscoloring of the potatoes which was assessed on an ordinal*

7 Generalized linear models

scale (A, B, C or D). The data are as follows:

Miscoloring	Treatment				Total
	a1b1	a1b2	a2b1	a2b2	
A	56	64	36	38	194
B	45	36	44	48	173
C	18	13	27	20	78
D	6	12	18	19	55
Total	125	125	125	125	500

A. The χ^2 test does not use the fact that the design is a two-factor design: it only gives as conclusion that there is a relation between treatment combination and response. Make a log-linear model analysis of these data, where the treatments are subdivided into factor A, factor B and possible interaction. Regard miscoloring as a qualitative variable, i.e. do not include it as ordinal.

B. Miscoloring, in fact, is an ordinal variable. Make a "proportional-odds" analysis of the same data that uses this fact.

For both A and B: Use a suitable computer program for the analysis.

Exercise 7.4 We study the survival of two species of snails. Groups of $n=20$ snails were kept under laboratory conditions during 1, 2, 3 or 4 weeks. New groups of snails were used for each number of weeks so the design is NOT "repeated measures". At the end of the period, it was counted how many of the snails that had died. Temperature (three levels) and humidity (four levels) were varied in the lab: Thus, we have in total (2 species) x (4 times) x (3 temperatures) x (4 levels of humidity) = 96 rounds, each with 20 snails. The data matrix has structure:

y	n	Species	Time	Temp	Humidity
4	20	0	1	1	1
5	20	0	1	1	2
2	20	0	1	1	3
...	...				

The data was analyzed using some program. Two models were used
Model 0:   y = species time temp humidity / distribution=binomial;
Model 1:   y = species time temp humidity species*time species*temp species*humidity / distribution=binomial;
Parts of the results:

Model 0:

	Estimate	Std error	z	p
Intercept	-1.405	0.971	-1.448	0.148
Species	1.309	0.163	8.005	<0.001
Time	1.503	0.102	14.693	<0.001
Temp	0.094	0.019	4.881	<0.001
Humidity	-0.107	0.0139	-7.700	<0.001

Null deviance: 539.72 on 95 degrees of freedom
Deviance: 55.07 on 91 degrees of freedom, AIC=223.93

Model 1:

	Estimate	Std error	z	p
Intercept	-2.505	1.646	-1.522	0.128
Species	3.055	2.039	1.499	0.134
Time	1.463	0.182	8.043	<0.001
Temp	0.096	0.032	3.023	0.003
Humidity	-0.089	0.023	-3.930	<0.001
Species*time	0.068	0.221	0.309	0.758
Species*humidity	-0.029	0.029	-1.018	0.309
species*temp	-0.003	0.040	-0.074	0.941

Null deviance: 539.72 on 95 degrees of freedom
Deviance   53.99 on 88 degrees of freedom, AIC=228.85

A. Write down formulas for the models that have been used: linear predictor, link and distribution.

B. Choose between the two models.
  i) based on a formal test
  ii) based on some suitable measure of model fit

C. Give a 95% confidence interval for the parameter that gives the effect of time, based on your "best" model in B.

D. According to Model 0, is there any significant different difference in survival between the two snail species?

E According to Model 0, which of the species has highest survival? (Note that the variable y gives the number of dead snails).

F. for fixed values of Time, Temp and Humidity, what is (according to model 0) the odds ratio for comparing species 0 with species 1?

G. the variables Time, Temp and Humidity can in principle be analyzed in two ways: either as numeric variables, or as "class" variables. According to the results, which of these ways has been used?

Chapter 8

Introduction to repeated-measures data

8.1 An example

Repeated-measures data, or longitudinal data, are data where the same measurements are made at several points in time or space on the same individuals. In many applications, the individuals have been randomized to different treatments, and one purpose may be to compare the time development between treatments. An example of repeated-measures data is given below; in later sections we will discuss possible analyses of this type of data.

Example 8.1 *The data in Table 8.1 come from an experiment that was performed in order to investigate how the growth of rats depends on different treatments. We will here use only two of the treatments ("Tr"): 1=Control; 3=Thiouracil. Weight measurements were made during five weeks: t=0 (birth weight, measured before the experiment started) and t=1, 2, 3 and 4 weeks.*

A typical feature of repeated-measures data is that the observations at different time points are correlated within individual. The analysis must take this into account. Thus, it is not possible to do a two-factor ANOVA with treatment and time as factors: this analysis would require that the observations at different time points are independent, which is not realistic.

8 Introduction to repeated-measures data

Table 8.1: Weight measurements at age 0, 1, 2, 3 and 4 weeks for rats randomized to two treatments.

Trt	Ind	T=0	T=1	T=2	T=3	T=4
1	1	57	86	114	139	172
1	2	60	93	123	146	177
1	3	52	77	111	144	185
1	4	49	67	100	129	164
1	5	56	81	104	121	151
1	6	46	70	102	131	153
1	7	51	71	94	110	141
1	8	63	91	112	130	154
1	9	49	67	90	112	140
1	10	57	82	110	139	169
3	18	61	86	109	120	129
3	19	59	80	101	111	126
3	20	53	79	100	106	133
3	21	59	88	100	111	122
3	22	51	75	101	123	140
3	23	51	75	92	100	119
3	24	56	78	95	103	108
3	25	58	69	93	114	138
3	26	46	61	78	90	107
3	27	53	72	89	104	122

8.2 Graphical displays for repeated-measures data

An important part of the analysis of repeated-measures data (and, indeed, of any kind of data) is to be able to visualize the data in a form that highlights the important patterns. Diggle et al (1994) give the following guidelines for graphing repeated-measures data:

1. Show as much of the relevant data as possible rather than data summaries.

2. Highlight aggregate patterns of potential scientific interest.

3. Identify both cross-sectional and longitudinal patterns in the data.

4. Make easy the identification of unusual individuals or unusual observations.

Graphical displays for repeated-measures data

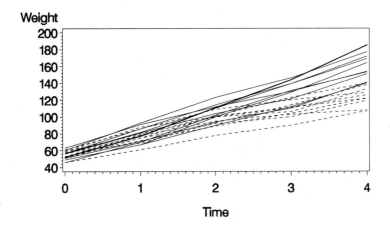

Figure 8.1: Individual profile plot for 20 rats under two treatments: treatment 1 (solid lines) and treatment 3 (dashed lines).

In the following we will give examples of a few useful types of graphs for repeated-measures data.

8.2.1 Individual profile plot

An individual profile plot shows the pattern of development over time for each individual. It is obtained by joining the data points for each individual. Figure 8.1 displays an individual profile plot for the data in Example 8.1.

8.2.2 Mean profile plot

The mean profile plot is a plot that shows the relation between "time" and the mean values for the different groups, treatments etc. in the study. The plot may also contain some indication of spread, for example the standard deviations. If the number of observations is not too large, the individual values may also be included. An example of a mean profile plot for the data in Example 8.1 is given in Figure 8.2.

8 Introduction to repeated-measures data

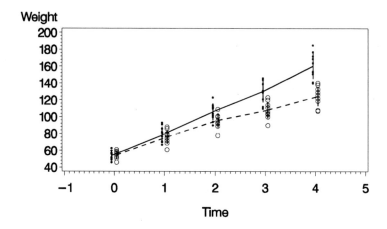

Figure 8.2: Mean profile plot for the growth pattern of rats under two treatments. Treatment 1 (solid dots) and treatment 3 (circles). A line indicating $\pm s$ is indicated at each time point.

8.2.3 Boxplots

A boxplot summarizes a set of data as a box that encloses the first, second and third quartile. The maximum and minimum values are also displayed. A boxplot of the data from Example 8.1 is presented in Figure 8.3. A boxplot often give a reasonable level of detail for summarizing the data: it indicates the trend at the same time as it shows the amount of variation in the data. For the boxplot in Figure 8.3, the time points were shifted somewhat to avoid that the boxes for the different treatments overlap.

8.3 Historical approaches

The analysis of repeated-measures data has attracted the attention of statisticians for a long time. Many different approaches have been suggested. In this section we will discuss three approaches that have often been applied, but which have now largely been replaced with other methods:

The "summary-measures" approach

Repeated-measures ANOVA

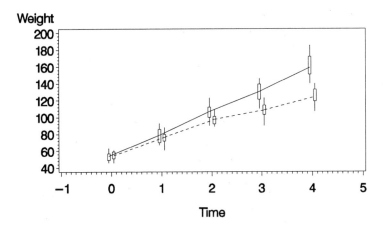

Figure 8.3: Box plot of data on growth of rats under two treaments. Group 1 (solid line) and group 3 (dotted line).

Growth-curve analysis using multivariate analysis of variance (MANOVA).

In Section 8.4 (page 194) we will introduce our preferred approach, which uses mixed linear models for analysis of this kind of data.

8.3.1 The "summary-measures" approach

A simple way of analyzing repeated-measures data is to proceed as follows:

1. Summarize the data for each individual as a small number of summary measures.

2. Analyze the summary measures using univariate methods such as t tests or ANOVA

In this approach, it is important that the summary measures really summarize the data in a good way. Some summary measures that are often used are

The change in y during the whole experiment, i.e. $y_{last} - y_{first}$.

The slope in a linear regression of y on time, for each individual.

8 Introduction to repeated-measures data

If the relation between y and time seems non-linear, it may be necessary to analyze also second- (or higher-) degree terms.

The time development can be described by a curve for each individual. The area under this curve is sometimes used as summary measure.

We will do two simple analyses of the data in Example 8.1 to compare the growth of the rats in the two treatment groups using the summary-measures approach.

The task is to find a suitable summary statistic and analyze this. We will use two summary measures. The first of these is the change in weight $(w_4 - w_0)$ during the whole experiment. The second alternative is to use the slope in a linear regression of weight on time, for each individual.

Analysis of total growth

The total growth $y_4 - y_0$ for the individuals in the two groups, for the data in Table 8.1, are :

```
Group 1   115   117   133   115   95   107   90   91   91   112
Group 3    68    67    80    63   89    68   52   80   61    69
```

We now have two samples, each of size $n = 10$. The hypothesis that the average total growth is equal for the two groups can be tested using a two-sample t test, or as a one-factor ANOVA The `ttest` procedure in SAS gives the following output:

```
                       The TTEST Procedure

                            Statistics

                                 Lower CL           Upper CL   Lower CL
Variable   Trt            N      Mean      Mean     Mean       Std Dev   Std Dev

Growth                1   10     96.282    106.6    116.92     9.9212    14.424
Growth                3   10     62.039     69.7     77.361    7.3659    10.709
Growth     Diff (1-2)            24.965     36.9     48.835    9.5984    12.703

                            Statistics
                                 Upper CL
           Variable   Trt        Std Dev    Std Err   Minimum   Maximum

           Growth             1   26.332    4.5612      90       133
           Growth             3   19.55     3.3864      52        89
           Growth   Diff (1-2)    18.785    5.6809
                              T-Tests

           Variable   Method           Variances    DF   t Value   Pr > |t|

           Growth     Pooled           Equal        18    6.50     <.0001
           Growth     Satterthwaite    Unequal      16.6  6.50     <.0001
```

We can conclude that the two treatments seem to differ in terms of average growth ($p < 0.0001$).

Analysis of the regression slopes

If we choose to use the individual regression slopes as summary statistic, we should:

1. Calculate the slope in the regression of weight on week for each individual

2. Use these slopes as the individual's data in an ANOVA or t test.

Note that, for these data, the regression slopes can be interpreted as "average weight change during one week of the experiment". The slopes thus have a substantive interpretation. Calculation of the individual regression lines can be automated in SAS. The slopes for the 20 rats are:

Group										
1	28.2	28.7	33.3	29.2	23.0	27.5	21.9	22.1	22.7	28.1
3	17.0	16.5	18.7	14.9	22.6	16.1	12.9	20.5	15.1	17.0

The slopes can be analyzed using e.g. a t test or ANOVA This time we use ANOVA The GLM procedure of SAS gives the following results.

```
                               The GLM Procedure

Dependent Variable: Estimate   Parameter Estimate

                                    Sum of
Source                     DF      Squares      Mean Square    F Value    Pr > F

Model                       1     437.1125000    437.1125000     38.46    <.0001

Error                      18     204.5970000     11.3665000

Corrected Total            19     641.7095000

                R-Square    Coeff Var      Root MSE    Estimate Mean

                0.681169    15.46170       3.371424        21.80500


Source                     DF    Type III SS   Mean Square    F Value    Pr > F

Trt                         1    437.1125000   437.1125000     38.46    <.0001
```

We can note that the average slopes seem to be different for the two treatments ($p < 0.0001$).

8.3.2 Repeated-measures ANOVA

The general linear models part of many statistical program packages often contain options for "repeated-measures". We will here consider

8 Introduction to repeated-measures data

how these options work, which underlying assumptions are used, and what kinds of results that can be expected from this type of analysis.

The model underlying a repeated-measures analysis of our example data may be written as

$$y_{ijk} = \mu + a_{ij} + \beta_j + \tau_k + (\beta\tau)_{jk} + e_{ijk}.$$

Here, μ is a general mean value, a_{ij} is the random effect of individual i within treatment j, β_j is the effect of treatment j, τ_k is the effect of time point k, $(\beta\tau)_{jk}$ is an interaction between treatment and time, allowing for different time developments for the different treatments, and e_{ijk} is a residual. Note that this model has many formal similarities with a split-plot model in ANOVA The individuals are the "main plots", and the measurements at different times are the "subplots" within individual. Therefore, the type of analysis we will describe is often called a split-plot approach.

If the model is correct, it can be shown that the covariance between two observations y_{ijk} and $y_{ijk'}$ on the same individual i, is

$$cov(y_{ijk}, y_{ijk'}) = \sigma_a^2,$$

the covariance between two observations at the same time for the same individual (which is, of course, the variance of an observation) is

$$cov(y_{ijk}, y_{ijk}) = Var(y_{ij}) = \sigma_a^2 + \sigma^2$$

and the covariance between any two observations on different individuals $y_{i'jk}$ and y_{ijk} is zero. This means that the covariance matrix between time points within individual is

$$\Sigma = \begin{pmatrix} \sigma^2 + \sigma_a^2 & \sigma_a^2 & \cdots & \sigma_a^2 \\ \sigma_a^2 & \sigma^2 + \sigma_a^2 & & \sigma_a^2 \\ \vdots & & \ddots & \sigma_a^2 \\ \sigma_a^2 & \sigma_a^2 & \sigma_a^2 & \sigma^2 + \sigma_a^2 \end{pmatrix}.$$

This corresponds to a correlation matrix with structure

$$\Sigma = \begin{pmatrix} 1 & \rho & \cdots & \rho \\ \rho & 1 & & \rho \\ \vdots & & \ddots & \vdots \\ \rho & \rho & \rho & 1 \end{pmatrix},$$

i.e. the correlations between measurements at different time points are

Historical approaches

assumed to be the same, regardless of the time distance between the measurements.

This structure is called a *compound symmetric* covariance structure. It can be shown that the usual ANOVA tests are valid under compound symmetry. In fact, the tests are valid under the more general condition of sphericity, which means that $Var(y_{ijk} - y_{ijk'})$ is constant. Programs for repeated-measures ANOVA may include an option to test the hypothesis that a compound symmetric structure is reasonable. They also often contain features to adjust the p values of the test if sphericity does not hold.

However, the compound symmetric covariance structure is rather unrealistic in practice. It may often be more realistic to use a covariance structure where the covariance (or correlation) depends on the size of the time interval between observations. We will return to this issue later.

A repeated-measures ANOVA of the data in Example 8.1 can be obtained using the following program. This program also produces output for the multivariate approach, to be discussed later.

```
PROC GLM data=growth;
  CLASS trt;
  MODEL t0 t1 t2 t3 t4 = trt;
REPEATED time polynomial / summary printm printe;
RUN;
```

This program will produce a separate ANOVA for each time point. In addition, the following results are obtained:

```
                    The GLM Procedure
           Repeated Measures Analysis of Variance
       Tests of Hypotheses for Between Subjects Effects

Source              DF    Type III SS    Mean Square   F Value   Pr > F

Trt                  1    4872.040000    4872.040000     14.23   0.0014
Error               18    6160.800000     342.266667
```

This gives the "split-plot" analysis of the over-all treatment ("trt") effect. We also get univariate tests of the effects of time, and the time*trt interaction:

```
                    The GLM Procedure
           Repeated Measures Analysis of Variance
     Univariate Tests of Hypotheses for Within Subject Effects

Source              DF    Type III SS    Mean Square   F Value   Pr > F

time                 4    95239.34000    23809.83500    627.31   <.0001
time*Trt             4     4625.06000     1156.26500     30.46   <.0001
Error(time)         72     2732.80000       37.95556
```

The assumption of sphericity can be tested (the `PRINTE` option in the program). The result is

8 Introduction to repeated-measures data

```
                     Sphericity Tests

                             Mauchly's
Variables             DF    Criterion     Chi-Square    Pr > ChiSq

Transformed Variates   9    0.0202173      64.045006      <.0001
Orthogonal Components  9    0.0202173      64.045006      <.0001
```

In this case, the departure from sphericity is clearly significant ($p < 0.0001$). This can be compensated for by using two different criteria, one by Greenhouse-Geisser and the other by Hyunh-Feldt. The tests are adjusted for non-sphericity and the corresponding p values are reported. In this case, the p values are the same as for the unadjusted test.

```
                         Adj Pr > F
Source                 G - G     H - F

time                   <.0001    <.0001
time*Trt               <.0001    <.0001
Error(time)


Greenhouse-Geisser Epsilon     0.3821
Huynh-Feldt Epsilon            0.4335
```

In this case, the results are so clear that most methods give similar conclusions.

Let us close the discussion on "repeated-measures ANOVA" by a quote. Diggle, Liang and Zeger (1994, p. 130) state, quite strongly, that "In summary, whilst ANOVA methods are undoubtedly useful in particular circumstances, they do not constitute a generally viable approach to longitudinal data analysis".

8.3.3 Growth curve analysis using MANOVA

Multivariate analysis of variance, MANOVA, is a generalization of analysis of variance to cases where there are several response variables for each individual. In the case of repeated-measures, the response variables are the measurements at the different time points. Thus, MANOVA considers the case where the data for each individual i consists of a response vector

$$\mathbf{y}_i = \begin{pmatrix} y_{i1} \\ y_{i2} \\ \vdots \\ y_{ip} \end{pmatrix}.$$

For a general discussion about MANOVA, see e.g. Johnson and Wichern (1992).

We will not consider in detail how the different tests in MANOVA are actually conducted. However, many different hypotheses can be tested within a MANOVA model. In the context of repeated-measures data, the following three hypotheses are often of interest (Davis, 2002):

1. H_{01}: The profiles for the different treatment groups are parallel (i.e. there is no treatment by time interaction).

2. H_{02}: There are no differences between treatment groups.

3. H_{03}: There are no differences between time points.

The way of testing the second and third of these hypotheses depends on the results of the first test.

The SAS program on page 191 will produce output that relates to the MANOVA approach. We report these below, in the order suggested in the list of hypotheses. First the test of parallelism:

```
              MANOVA Test Criteria and Exact F Statistics
                for the Hypothesis of no time*Trt Effect
                  H = Type III SSCP Matrix for time*Trt
                         E = Error SSCP Matrix

                         S=1      M=1       N=6.5

Statistic                      Value       F Value    Num DF    Den DF    Pr > F

Wilks' Lambda                  0.26831525   10.23      4         15        0.0003
Pillai's Trace                 0.73168475   10.23      4         15        0.0003
Hotelling-Lawley Trace         2.72695917   10.23      4         15        0.0003
Roy's Greatest Root            2.72695917   10.23      4         15        0.0003
```

The test of the treatment effect is the same as reported above:

```
                         The GLM Procedure
                   Repeated Measures Analysis of Variance
             Tests of Hypotheses for Between Subjects Effects

Source            DF      Type III SS      Mean Square    F Value    Pr > F

Trt               1       4872.040000      4872.040000    14.23      0.0014
Error             18      6160.800000      342.266667
```

Finally, the MANOVA test of the time effect:

```
MANOVA Test Criteria and Exact F Statistics for the Hypothesis of no time Effect
                      H = Type III SSCP Matrix for time
                         E = Error SSCP Matrix

                         S=1      M=1       N=6.5

Statistic                      Value        F Value    Num DF    Den DF    Pr > F

Wilks' Lambda                  0.01432068   258.11     4         15        <.0001
Pillai's Trace                 0.98567932   258.11     4         15        <.0001
Hotelling-Lawley Trace         68.82907503  258.11     4         15        <.0001
Roy's Greatest Root            68.82907503  258.11     4         15        <.0001
```

The MANOVA approach to modeling repeated-measures data has several advantages. First, in contrast to the "repeated-measures ANOVA" approach, the correlations between observations within individual are left free to vary. This gives a more realistic model for the dependence between observations over time. Second, the MANOVA approach permits tests of a number of hypotheses that are of interest for this kind of data.

But there are disadvantages too:

The data must be complete. If an observation is missing at a single time point for an individual, all data for this individual are deleted.

The time points must be the same for all individuals.

These two disadvantages may be rather serious. It is very common that individuals get occasional "holes" in their data. Deletion of all individuals where some data are missing may sometimes considerably reduce the sample size. For example, if there are 5 time points, and the probability that an observation is missing is 5%, we would expect that 23% of the individuals have missing data on at least one time point. Also, the requirement that all individuals have observations at the same time points is often difficult to adhere to in practice. Both these problems can be handled within the framework of mixed linear models.

8.4 Analysis as a mixed linear model

8.4.1 Introduction to mixed models

Mixed models are statistical models where the levels of some of the factors in the model are fixed while other factors are regarded as randomly selected from a population of levels. A model that contains only fixed effects is a fixed-effects model; if only random effects are included the model is a random-effects model; and if some effects are fixed and some are random the model is called a mixed model. In repeated-measures models, the individual can often be regarded as a random factor.

A typical feature of mixed models is that they contain more than one "error term". That is, the model parameters include more than one variance component. In fixed models, there is only one error term, SS_e, which corresponds to one model parameter, σ_e^2. For example, in a mixed model with "individual" as random, the variance between individuals is a second variance component.

Formally, a mixed model can be written in matrix terms as

$$\mathbf{y} = \mathbf{XB} + \mathbf{Zu} + \mathbf{e}.$$

Here, $\mathbf{y}$ is a vector that contains the responses for all individuals, stacked on top of each other. $\mathbf{X}$ is a design matrix, and $\mathbf{B}$ is a parameter vector, as in general linear models. As regards the random effects, $\mathbf{u}$ is a vector of random effects and $\mathbf{Z}$ is a design matrix for the random effects. Finally, $\mathbf{e}$ is a vector of residuals.

The parameters to be estimated are the fixed parameters in **B**, the covariance matrix $\mathbf{R} = Cov(\mathbf{u})$ and the covariance matrix $\mathbf{G} = Cov(\mathbf{e})$. The two sets of random effects **u** and **e** are often referred to as the R-side and G-side effects, respectively. The G-side effects are sometimes described as "within subject" effects.

Estimation of parameters in this type of models is often made using methods such as REML (restricted maximum likelihood). This gives estimates of the fixed parameters **B**, and of the variances and covariances of random effects, along with estimated standard errors of the estimates.

Thus, to specify a mixed model you need to give one linear model for the fixed effects and another model for the random effects. In the SAS procedure MIXED, this is done as two different model statements. In some other software, all effects, fixed as well as random, are included in the model, while some other statements indicates which of the factors are random.

8.4.2 A model with compound symmetry

We will first make an analysis that is similar to the "repeated-measures ANOVA" model, discussed earlier. For analysis with the Mixed procedure, the data should be re-organized as follows:

Obs	Trt	Ind	T0	t	w
1	1	1	57	0	57
2	1	1	57	1	86
3	1	1	57	2	114
4	1	1	57	3	139
5	1	1	57	4	172
6	1	2	60	0	60
7	1	2	60	1	93
8	1	2	60	2	123
9	1	2	60	3	146
10	1	2	60	4	177
11	1	3	52	0	52

This means that we list the data as one line for each time point, i.e. the data for each individual are listed on several lines.

A SAS program to analyze the data is as follows:
```
PROC MIXED data=growth2;
   CLASS trt ind t;
   MODEL w = trt t ;
   RANDOM ind*trt ;
RUN;
```
The program assumes that treatment, time (t) and the code for individual are class variables. In later analyses, time may be modeled as a numeric variable using linear, quadratic and cubic trends. The RANDOM

8 Introduction to repeated-measures data

statement indicates that the individuals (here coded as `ind*trt`) are regarded as random.

The output from this program is as follows:

```
              The Mixed Procedure

              Covariance Parameter
                   Estimates

              Cov Parm        Estimate

              Trt*Ind          49.0905
              Residual         96.8139
                 Fit Statistics

      -2 Res Log Likelihood           737.5
      AIC  (smaller is better)        741.5
      AICC (smaller is better)        741.7
      BIC  (smaller is better)        743.5
            Type 3 Tests of Fixed Effects

                     Num    Den
      Effect          DF     DF    F Value    Pr > F

      Trt              1     18     14.23     0.0014
      t                4     76    245.93     <.0001
```

The results indicate that the treatment differences are significant; we also find a significant effect of time.

It may be of interest to include in the model the Trt*t interaction. A similar model as above, but including that interaction, gives the following results:

```
            Type 3 Tests of Fixed Effects

                     Num    Den
      Effect          DF     DF    F Value    Pr > F

      Trt              1     18     14.23     0.0014
      t                4     72    627.31     <.0001
      Trt*t            4     72     30.46     <.0001
```

Note that all F values and p values for treatment, time and treatment*time are the same as in the "repeated-measures ANOVA" analysis on page 191. We find that the t*trt interaction is highly significant. This indicates that the pattern of growth is different for the two treatments; the average growth curves are not parallel.

The approach to modeling the data used in this example is based on the assumption that the correlation structure over time is compound symmetric. This structure is often not realistic for repeated-measures data: it means that the correlation between measurements is the same, regardless of the distance in time between measurements. A more realistic structure of the correlations is needed.

Table 8.2: Some correlation patterns useful for repeated-measures data.

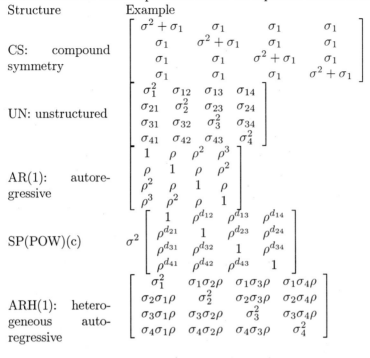

8.4.3 Modeling the covariance structure

The covariance matrix within individual can be modeled using the "RE-PEATED" statement in Proc MIXED. There are several options for selecting the structure. Table 8.2 lists a few of the most commonly used options. For a complete account, see the SAS (2008) manual.

Some comments about the use of these structures for repeated-measures data are:

- CS: The compound symmetric structure mimics the "repeated-measures ANOVA" approach. This is not realistic but may sometimes be used as an approximation if other alternatives fail.

- UN: unstructured. This is the most general structure and puts no restrictions on the covariances. It corresponds to the structure that is implicit in MANOVA. One drawback is that many parameters need to be estimated. This means that the program may not converge when this structure is used, if the data are sparse. Then, a simpler model should be specified.

- AR(1): This autoregressive structure assumes that the correlations between observations decrease with increasing time span. This is often a realistic structure. It includes rather few parameters, but it assumes that the time points are equidistant.

- SP(POW): The Spatial Power structure also assumes that the correlations decrease with increasing time span. It is a realistic alternative for data which are not equidistant in time.

- ARH(1): This is similar to AR(1) but with more parameters, allowing for different variances at the different time points.

8.4.4 Using the "repeated" statement

The "repeated" statement is used in Proc MIXED to indicate the type of covariance structure to use. A program to do this is as follows.

```
PROC MIXED data=growth2;
  CLASS trt ind t;
  MODEL w = trt t trt*t ;
  REPEATED /subject=ind*trt type=un;
RUN;
```

The REPEATED statement identifies the subjects within which observations are repeated. In this case the subjects are individuals within treatments. The correlation structure was given as UN, i.e. as unstructured. This means that all correlations/covariances are free to be estimated. This may not work in all data sets. If the analysis using an unstructured covariance matrix does not work, the AR(1) structure can be used for equidistant measurements, and the spatial power structure for non-equidistant measurements. The exponent d would then be set to some function of the distance in time between measurements.

Some output based on the unstructured covariance model is as follows. First, we get an estimate of the covariance matrix of dimension 5×5:

Analysis as a mixed linear model

```
Covariance Parameter Estimates

Cov Parm     Subject      Estimate

UN(1,1)      Trt*Ind       25.7833
UN(2,1)      Trt*Ind       38.7167
UN(2,2)      Trt*Ind       77.8111
UN(3,1)      Trt*Ind       34.6333
UN(3,2)      Trt*Ind       71.4778
UN(3,3)      Trt*Ind       85.3111
UN(4,1)      Trt*Ind       28.1444
UN(4,2)      Trt*Ind       57.5500
UN(4,3)      Trt*Ind       89.9667
UN(4,4)      Trt*Ind      126.47
UN(5,1)      Trt*Ind       21.5667
UN(5,2)      Trt*Ind       44.4889
UN(5,3)      Trt*Ind       87.6556
UN(5,4)      Trt*Ind      134.42
UN(5,5)      Trt*Ind      178.71
```

Next, we get a section describing the fit of the model to data. This can be used to compare competing models; see page 202.

```
              Fit Statistics

-2 Res Log Likelihood          551.8
AIC  (smaller is better)       581.8
AICC (smaller is better)       588.3
BIC  (smaller is better)       596.8


    Null Model Likelihood Ratio Test

    DF      Chi-Square     Pr > ChiSq

    14        139.99         <.0001
```

Finally, we obtain the tests of the fixed effects of the model.

```
       Type 3 Tests of Fixed Effects

           Num      Den
Effect     DF       DF      F Value    Pr > F

Trt         1       18       14.23     0.0014
t           4       18      309.73     <.0001
Trt*t       4       18       12.27     <.0001
```

8.4.5 Further improvements of the model

Using baseline measurements

In many repeated-measures applications, one baseline measurement was taken before the treatments started. This is the case in the example used in this chapter. It is not realistic to assume that the treatments have an effect on this measurement. On the other hand, the baseline measurement can certainly be assumed to be related to subsequent measurements on the same individual, i.e. the baseline measurements contain valuable information.

A good way of using such measurements is to delete them from the

8 Introduction to repeated-measures data

list of "dependent" variables, but to include them as covariates in the analysis. This means that the response (y) is assumed to be related to treatment, time, possible interactions, and the baseline value. This type of model, a so called covariance analysis model, would make the analysis considerably stronger.

We analyzed the example data using this approach. Time 0 was deleted from the analysis ("WHERE t>0"). The weight at time 0 (T0) was used as a covariate. The program was:

```
PROC MIXED data=growth2;
  CLASS trt ind t;
  MODEL w = trt t trt*t t0;
  REPEATED /subject=ind*trt type=un;
  WHERE t>0;
RUN;
```

Some of the output was:

```
                    Fit Statistics

         -2 Res Log Likelihood          440.4
         AIC (smaller is better)        460.4
         AICC (smaller is better)       464.1
         BIC (smaller is better)        470.4


             Null Model Likelihood Ratio Test

              DF     Chi-Square      Pr > ChiSq

               9         99.44          <.0001
```

Type 3 Tests of Fixed Effects

Effect	Num DF	Den DF	F Value	Pr > F
Trt	1	17	30.95	<.0001
t	3	17	213.72	<.0001
Trt*t	3	17	12.78	0.0001
T0	1	17	50.62	<.0001

As we can see, the effect of the initial weight T0 is highly significant. This means that we have removed the effect of initial weight, thus reducing the influence of some of the variation in the data.

Polynomials in time

If there are t time points, the time effect will use $t-1$ degrees of freedom if it is included as a "class" variable. In addition, each interaction with time will use $k \cdot (t-1)$ df, where k is the number of d.f. for the factor interacting with time. It is sometimes possible to simplify the model by approximating the time trend with polynomials. A polynomial model of degree $t-1$ is formally equivalent to a model with time as a class variable. If the degree of the polynomial is smaller than $t-1$, we have a simpler model than the original one.

Analysis as a mixed linear model

Some outputs of results for our example data, using polynomials of degree 1, 2 and 3, are as follows:

Degree 1:

Type 3 Tests of Fixed Effects

Effect	Num DF	Den DF	F Value	Pr > F
Trt	1	18	1.09	0.3111
t	1	18	1200.51	<.0001
t*Trt	1	18	36.58	<.0001

Degree 2:

Type 3 Tests of Fixed Effects

Effect	Num DF	Den DF	F Value	Pr > F
Trt	1	18	0.01	0.9097
t	1	18	552.21	<.0001
t*Trt	1	18	0.82	0.3768
t*t	1	18	9.32	0.0068
t*t*Trt	1	18	12.21	0.0026

Degree 3:

Type 3 Tests of Fixed Effects

Effect	Num DF	Den DF	F Value	Pr > F
Trt	1	18	0.09	0.7644
t	1	18	160.48	<.0001
t*Trt	1	18	0.00	0.9965
t*t	1	18	1.85	0.1906
t*t*Trt	1	18	1.45	0.2438
t*t*t	1	18	0.62	0.4414
t*t*t*Trt	1	18	0.29	0.5939

As we can see, the third-degree terms are far from being significant. This suggests that the time trend can be approximated by a second degree polynomial. This is also supported by the values of AIC which are 631.0 (degree 1); 619.1 (degree 2) and 620.1 (degree 3); AIC is smallest for degree 2. A discussion on the use of the AIC follows below.

201

8 Introduction to repeated-measures data

Model-building tools

As we have seen, there are many options when fitting models for repeated-measures data. Both the fixed part and the random part of the model may be modified by adding or deleting factors. In addition, the covariance structure may be modeled in different ways. Formal and informal tools for model building are available, and we will briefly describe them.

Formal tests Nested models are models where one of them is a subset of the other, in terms of the parameters included. When comparing nested models it is possible to formally test the hypothesis that the parameters that are included in the larger model, but not in the smaller, are all zero.

The test is constructed by comparing the values of $-2\log L$ for the two models. This value is often given in the computer printouts. It holds that
$$\chi^2 \approx (-2\log L_1) - (-2\log L_2),$$
in large samples, can be regarded as an approximate χ^2 statistic. The degrees of freedom is equal to the difference in number of parameters between the models. This can be used to test the hypothesis that the extra parameters in the larger model are all zero. The hypothesis is rejected if the observed test statistic is large.

Example 8.2 *For the data in Example 8.1, an analysis using an unstructured covariance matrix gives $-2logL = 440.4$. The AR(1) structure gives $-2logL = 481.7$. The unstructured model has 10 covariance parameters while the AR(1) model has 2. Thus, a test of the hypothesis that the extra covariance parameters are irrelevant is given by*
$$\chi^2 = 481.7 - 440.4 = 41.3$$
on $10 - 2 = 8$ degrees of freedom. This is highly significant ($p < 0.0001$). Thus, the model cannot be simplified to an AR(1) structure.

Informal test Akaike (1973) suggested a statistic, the Akaike information criterion (AIC), that can be used to compare competing models. The models need not be nested, but they should use the same data. (For example, the AIC should not be used to compare a model based on raw data with a model based on log-transformed data.) The statistic is based on comparing $(-2\log L)$, so that a model with more parameters gets "punished". Thus, the aim of the AIC is to obtain a simple model with as few parameters as possible. Among a set of competing models, the strategy would then be to choose the model that has the smallest value of AIC[1]. However, as in all model building, knowledge about the subject-matter area and other factors also need to be taken into account.

Example 8.3 *For the data in Example 8.1, the analysis using an unstructured covariance matrix, reported above, produced $AIC = 460.4$. A similar analysis but using an autoregressive covariance structure (AR(1)), has $AIC = 485.7$. Since a smaller AIC indicates a better fit, the unstructured covariance matrix is to be preferred.*

8.4.6 A strategy for mixed model analysis

A useful strategy for analysis of simple repeated-measures data is as follows. We assume that we want to study the effects of some fixed factors (treatments or strata). Each subject has been measured on a number of occasions. These occasions are the same for all subjects, except for occasional missing values.

1. **The fixed part:** State a baseline model for the fixed part of the model. This model should include treatments and other fixed factors including relevant interactions, time (modeled as a CLASS variable), and possible interactions between time and the fixed factors. Baseline values, i.e. values recorded before any treatments started, may be included as covariates.

2. **The "repeated" part:** State a REPEATED statement for time. Here, indicate the subjects within which the measurements are repeated. As a first attempt, use an unstructured covariance matrix within individual. Record model diagnostic statistics like the Akaike information criterion (AIC) for this model.

3. **Simplify the covariance structure:** If the program does not converge: simplify the model by changing the covariance structure. For equally spaced time points, you may want to try an AR(1) structure. If the time points are not equally spaced, a spatial power structure may be appropriate. Other alternatives are also possible (compound symmetry, etc.). This type of model simplification may be useful even if the first model worked from a technical point of view. Compare models based on i.a. the AIC.

4. **Simplify the fixed part:** Consider possible simplifications of the fixed part of the model. For this model simplification the model

---

[1] Some care is needed here. Some programs report the AIC with opposite sign. Then, a large value of AIC is better. SAS prints "smaller is better" when reporting AIC, in order to avoid this confusion.

building advice in regression and ANOVA may be used. Criteria for model simplification are the p values of the terms, and a comparison of AIC values.

5. **Polynomials in time:** The effect of time has so far been modeled as a CLASS variable. If there are t time points, it is possible to model the effect of time using a polynomial of degree up to $(t-1)$. One option towards model simplification is to model time as a polynomial of degree smaller than $(t-1)$. Note that if the original model contains interactions between fixed factors and time, the polynomial model should include interactions between the fixed factors and each power of the polynomial. That is, if the model with time as CLASS variable is Y = A T A*T; then a second degree polynomial model would include Y = A T T*T A*T A*T*T;. It would not be meaningful to assume e.g. an interaction with the second degree term but not with the first degree term. In cases where the functional form of the relation with time is specified by some theory, this should of course be used.

In this brief account we have assumed that the time points are essentially the same for all individuals. However, there are cases where the individuals are measured on different time points. An example is when patients receive checkups according to a timetable that suits both the patient and the clinic. In such cases, the "mixed-model" approach can still be used, but the time development has to be modeled either as a parametric function (like a polynomial), or using some nonparametric regression approach. Advanced texts, such as Fitzmaurice et al (2004) treat this problem.

8.5 Analysis of Cross-over data

8.5.1 Introduction

A cross-over study (or change-over study) is a study where the subjects change treatment during the course of the experiment. This gives a specific type of repeated-measures data. Cross-over designs are often efficient since the treatment comparisons are made within individual. On the other hand, they can also bring problems. One of the problems is that remaining treatment effects from one period may affect the result during a subsequent period (carry-over effects). This may have to be accounted for during the analysis.

Analysis of Cross-over data

Example 8.4 *We want to compare three treatments on a set of patients. The trial is subdivided into time periods and the patients change treatment between periods. It is common to arrange the patients in blocks such that the design, within each block, is a Latin square, and such that the complete set of Latin squares is balanced. A Latin square is a square arrangement of symbols where each symbol is present once in each column and once in each row. Example:*

A	B	C
C	A	B
B	C	A

The experiment might look like:

Block	1			2			...	4		
Patient	1	2	3	4	5	6	...	10	11	12
Period										
1	A_1	A_2	A_3	A_1	A_2	A_3		A_2	A_1	A_3
2	A_2	A_3	A_1	A_3	A_1	A_2		A_1	A_3	A_2
3	A_3	A_1	A_2	A_2	A_3	A_1		A_3	A_2	A_1

In this experiment the patients were blocked with three patients per block. The experiment is divided into three periods such that each patient tries each of the $a = 3$ treatments. Patient 1 gets treatment A_1 during period 1, treatment A_2 during period 2 and treatment A_3 during period 3. Wash-out periods are often inserted between the "active" periods in order to remove possible carry-over effects.

Start Period 1 wash-out Period 2 wash-out Period 3

8.5.2 Designs

The design of a crossover study is rather intricate, and depends on the number of treatments, the number of periods and the number of patients. see e.g. Jones and Kenward (2003). Cross-over studies can be described in terms of the *sequence* of the treatments. For example, in a study with two treatments and two periods the sequences can be

$$A\ B$$
$$B\ A$$
$$A\ A$$
$$B\ B$$

8 Introduction to repeated-measures data

Table 8.3: Data from a change-over experiment with eight Latin squares.

Square	Patient	Per1	Per2
1	1	5.1	3.8
1	2	2.9	3.9
2	1	0.6	1.0
2	2	1.6	2.3
3	1	4.8	3.1
3	2	4.0	5.8
4	1	4.4	4.9
4	2	1.6	0.8
5	1	2.3	1.3
5	2	4.1	4.7
6	1	4.9	2.3
6	2	3.2	0.9
7	1	6.8	4.5
7	2	2.3	4.0
8	1	6.1	2.2
8	2	3.4	3.6

However, it is common that only the first two sequences are used. The design of a crossover study can be described by listing the sequences of the treatments that are used in the study, for example AB/BA or $ABBA/BAAB$. Incomplete block designs, where each patient receives only some of the treatments, are also possible.

Example 8.5 *The data in Table 8.3 represent a two-period change-over experiment undertaken in eight Latin squares. The first patient in each square takes the treatments in order (A, B) while the second patient has the order (B, A).*

8.5.3 Mixed models

Following Jones and Kenward (2003), the observed value of the response can be decomposed as follows:

$$y_{ijk} = \mu + s_{ik} + \pi_j + \tau_{d[i,j]} + \lambda_{d[i,j-1]}$$

Here, μ is a general mean value; s_{ik} is a random effect of individual; π_j is a fixed effect of period j, $\tau_{d[i,j]}$ is a fixed effect of the treatment received by individual i during period j, and $\lambda_{d[i,j-1]}$ is a fixed carry-over effect of the treatment received by patient i during period $j-1$. The model is a mixed model with individual as a random effect. Depending on the

Analysis of Cross-over data

design, it is not always possible to estimate all parameters of the model: some effects may be confounded. Tools like the Mixed procedure of SAS can be used to analyze the data.

Example 8.6 *The data in Table 8.3 were analyzed using Proc Mixed. The data were given to the procedure in the following form:*

Obs	Square	Patient	y	trt	period	prev
1	1	1	5.1	A	1	none
2	1	1	3.8	B	2	A
3	1	2	2.9	B	1	none
4	1	2	3.9	A	2	B
5	2	1	0.6	A	1	none
6	2	1	1.0	B	2	A
7	2	2	1.6	B	1	none
8	2	2	2.3	A	2	B
9	3	1	4.8	A	1	none
10	3	1	3.1	B	2	A
11	3	2	4.0	B	1	none
12	3	2	5.8	A	2	B
13	4	1	4.4	A	1	none
14	4	1	4.9	B	2	A
15	4	2	1.6	B	1	none
16	4	2	0.8	A	2	B
17	5	1	2.3	A	1	none
18	5	1	1.3	B	2	A
19	5	2	4.1	B	1	none
20	5	2	4.7	A	2	B
21	6	1	4.9	A	1	none
22	6	1	2.3	B	2	A
23	6	2	3.2	B	1	none
24	6	2	0.9	A	2	B
25	7	1	6.8	A	1	none
26	7	1	4.5	B	2	A
27	7	2	2.3	B	1	none
28	7	2	4.0	A	2	B
29	8	1	6.1	A	1	none
30	8	1	2.2	B	2	A
31	8	2	3.4	B	1	none
32	8	2	3.6	A	2	B

The SAS program has the following structure:

```
PROC MIXED data=co;
CLASS square patient period trt prev;
MODEL y = trt period prev / htype=1 solution;
RANDOM patient*square;
RUN;
```

Part of the results were as follows:

Solution for Fixed Effects

Effect	trt	prev	period	Estimate	Standard Error	DF	t Value	Pr > |t|
Intercept				1.7625	1.1603	14	1.52	0.1510
trt	A			1.4875	0.8001	14	1.86	0.0841
trt	B			0	.	.	.	.
period			1	1.1250	0.8001	14	1.41	0.1815
period			2	0	.	.	.	.
prev		A		1.1250	1.4326	14	0.79	0.4454
prev		B		0	.	.	.	.
prev		none		0	.	.	.	.

Type 1 Tests of Fixed Effects

Effect	Num DF	Den DF	F Value	Pr > F
trt	1	14	6.74	0.0212
period	1	14	2.49	0.1368
prev	1	14	0.62	0.4454

8 Introduction to repeated-measures data

The treatment effect is significant (p = 0.0212)) while neither the period effect nor the carry-over effect is significant. In this example, we use a random patient effect. This gives a compound symmetric structure for the correlation between measurements within individual; this works fine here since there are only two time points In other examples, an unstructured (or e.g. AR(1)) structure might be used.

Mixed Generalized Linear Models

The mixed-model approach can also be used in models where the response is not normal. Some types of such data can be analyzed using Generalized linear models. In this case, the linear predictor is

$$\eta_{ijk} = \mu + s_{ik} + \pi_j + \tau_{d[i,j]} + \lambda_{d[i,j-1]}.$$

The distribution and the link function can be selected based on the type of data. This gives a generalized linear mixed model.

Example 8.7 *Thirty-three children between the ages of 6 and 16 years, all suffering from monosymptomatic nocturnal enuresis, were enrolled in a study. The study was carried out with a double-blind randomized three-period cross-over design. The children received 0.4 mg Desmopressin, 0.8 mg Desmopressin or placebo tablets at bedtime for five consecutive nights with each dosage. A wash-out period of at least 48 hours without any medication was interspersed between treatment periods. Wet and dry nights were documented; for more details about the study and its analysis see Neveus et al (1999), and Olsson and Neveus (2000). The data consisted of nightly recordings, where a dry night was recorded as 1 and a wet night as 0. The nights were grouped into sets of five nights where the same treatment had been given. The structure of the data is given in Table 8.4. Only one patient is listed; the original data set contained 33 patients.*

Generalized linear mixed models containing different combinations of model parameters were tested. The Glimmix procedure of the SAS package was used. The distribution was taken to be binomial with $n = 5$ and the link was logistic. The results are summarized in Table 8.5. The numbers in the table are p-values to assess the significance of the different factors. Patient was included as a random factor in all models.

Based on these results, it was concluded that a model containing a random patient effect and fixed effects of dose and period, provided an appropriate description of the data. Neither the sequence effect nor the after effect was anywhere close to being significant in any of the analyses.

Analysis of Cross-over data

Table 8.4: Raw data for one patient in the enuresis study

Patient	Period	Dose	Night	Dry
1	1	1	1	1
1	1	1	2	1
1	1	1	3	1
1	1	1	4	1
1	1	1	5	1
1	2	0	1	0
1	2	0	2	0
1	2	0	3	1
1	2	0	4	0
1	2	0	5	0
1	3	2	1	1
1	3	2	2	1
1	3	2	3	1
1	3	2	4	1
1	3	2	5	1

Table 8.5: p values for different models for the enuresis data

Effects included	Dose	Period	Sequence	After effect
Dose	.0001			
Dose, Seq	.0001		.7442	
Dose, Period	.0001	.0938		
Dose, After eff.	.0001			.6272
Dose, After eff., Period	.0001	.0759		.8713
Dose, After eff., Seq.	.0001		.7762	.6573
Dose, Period, Seq	.0001	.0898	.7577	

209

8 Introduction to repeated-measures data

Further analyses using pairwise comparisons revealed that there were no significant differences between doses but that the drug had a significant effect at both doses.

8.6 Exercises

Exercise 8.1 *A friend, who is a physiology student, has made an experiment on the effects of two operation methods on the level of stress hormones in rabbits. The experiment was designed as follows. Ten rabbits were randomized to the two operation methods with five rabbits to each. The stress hormone level was measured on each animal just after the operation (time 0) and 1, 2 and 3 hours thereafter.*

The analysis suggested by your friend is a two-factor ANOVA with operation method and time as qualitative factors. The symbolic SAS/Minitab model is

Stress = Method Time Method*Time;

A. The analysis suggested by your friend is not good. Explain to your friend why it is not.

B. If you still use the suggested model, explain what will probably happen to the p values, as compared to a more correct analysis.

C. Suggest a better analysis of the data.

Exercise 8.2 *The following analysis was published in a journal. Assume that the reported sums of squares and degrees of freedom are correct.*

We sampled leafminer density on four shaded and three control trees biweekly from June until August, at a total of five measurement occasions. A repeated-measures ANOVA model was used for analysis with densities as the dependent variable and treatment (sun/shade) and time (sample date) as factors.

Source	SS	df	F	p
Treat	1.020	1	17.65	<0.001
Trees	5.187	5	17.95	<0.001
Date	7.646	4	33.07	<0.001
Date x treat	0.349	4	1.51	>0.05
Error	1.156	20		
Total		34		

A. Something has gone wrong in the analysis: the ANOVA table does not give a correct "repeated-measures ANOVA". What is wrong?

B. Correct the error so that a correct "repeated-measures ANOVA" is obtained.

C. Even after correction, the analysis is not optimal. Explain why and suggest a better analysis of the data. (No new computations are needed).

Exercise 8.3 *Data for this question (Davis 2002 p. 99; Grizzle and Allen, 1969) are from an experiment studying the level of coronary sinus*

8 Introduction to repeated-measures data

potassium following coronary occlusion on dogs. There were four treatments with nine dogs for each treatment. The following treatments were used:

1. Untreated (Control).

2. Extrinsic cardiac denervation 3 weeks prior to the experiment.

3. Extrinsic cardiac denervation immediately before the experiment.

4. Bilateral thoracic sympathectomy and stellectomy 3 weeks prior to the experiment.

Measurements were made on each dog 1, 3, 5, 7, 9, 11, 13 minutes after occlusion.

Question:

Analyze these data to find possible treatment differences. Note that the data are of a repeated-measures type: use some method we have discussed in the course to account for this.

The following variables are used in the data sets:

Data set 1: one line per dog

Variable	Explanation
Treat	Treatment number (1—4 as listed above)
Dog	Dog number (1—36)
$y1, y3, \ldots, y13$	Potassium level at time 1, 3, ..., 13 minutes

Data set 2: one line per time point

Variable	Explanation
Treat	Treatment number (1—4 as listed above)
Dog	Dog number (1—36)
t	Time when measurement was taken
y	Potassium level

(Data are supplied on the home page for the book).

Chapter 9

Introduction to multivariate methods

9.1 Describing multivariate data

9.1.1 The data matrix

In multivariate statistics, the data are given as a *data matrix*. The data matrix is designed so that each row is one *"individual"*, and each column is one *"variable"*; see Table 9.1. Individual values in the data matrix can be written as x_{ij} where the first index indicates row (individual) and the second index indicates variable.

Table 9.1: A data matrix

Individual	Variable			
	1	2	...	p
1	x_{11}	x_{12}		x_{1p}
2	x_{21}	x_{22}		x_{2p}
...			⋱	
n	x_{n1}	x_{n2}		x_{np}

Example 9.1 *Table 9.2 is a data matrix that contains chemical measurements for a number of soil samples. In this example, the soil samples are the "individuals" and the different chemical measurements (pH etc.) are the variables.*

9 Introduction to multivariate methods

Table 9.2: Results of chemical analyses of a number of soil samples.

pH	EC	DOC	Cd	Cr	Cu	Ni	Pb	Zn
6.71	328.30	52.21	2.80	10.10	254.00	17.60	14.80	0.61
6.35	294.70	51.69	3.60	10.20	175.00	17.10	20.50	0.66
6.51	304.20	51.07	1.80	9.00	177.00	14.30	17.30	0.66
5.94	144.50	32.52	0.35	8.40	143.00	7.40	21.00	0.44
5.90	140.50	27.97	0.32	7.30	119.00	5.50	20.70	0.43
5.92	144.60	27.76	0.33	6.90	123.00	6.40	19.90	0.43
5.61	133.30	18.26	0.46	4.80	81.50	7.50	15.70	0.53
5.60	133.90	16.54	0.46	4.40	71.50	7.30	12.60	0.58
5.66	136.20	17.57	0.44	3.90	71.50	7.20	13.00	0.63
5.51	140.10	14.63	0.55	3.98	74.00	7.90	10.00	0.73
5.42	136.90	14.18	0.52	4.07	63.50	7.00	11.20	0.72
5.53	137.90	14.88	0.53	3.57	69.00	7.80	10.20	0.78
5.41	128.00	12.99	0.63	3.51	71.00	7.60	11.90	0.78
5.36	126.20	12.93	0.55	3.74	75.00	9.50	11.70	0.76
5.41	128.20	12.69	0.58	3.00	84.50	7.70	9.10	0.74

9.1.2 Uses of multivariate statistics

The purpose of the analysis may be, for example:

- Simplification. If some variables are highly related we can replace them with an index: it is easier to handle one (or a few) variables than to handle many.

- Sorting, grouping. Can the individuals be classified into different types?

- Study dependence between the variables. For example, is there any relation between the weight of a patient and the cholesterol level?

- Prediction. Can some combination of variables be used to predict if the patient will improve?

- Construction and testing of hypotheses.

9.1.3 The mean vector and the covariance matrix

A data matrix can be summarized in the form of descriptive statistics. The mean value for variable x_j is

Describing multivariate data

$$\overline{x}_j = \frac{1}{n}\sum_{i=1}^{n} x_{ij}$$

The mean values of the different variables can be compiled into a *mean vector*:

$$\overline{x} = \begin{bmatrix} \overline{x}_1 \\ \overline{x}_2 \\ \vdots \\ \overline{x}_p \end{bmatrix}$$

The *variance* of variable x_j is

$$s_j^2 = s_{jj} = \frac{1}{n-1}\sum_{i=1}^{n}(x_{ij} - \overline{x}_j)^2.$$

The *covariance* between two variables x_j and x_k is defined as

$$s_{jk} = \frac{1}{n-1}\sum_{i=1}^{n}(x_{ij} - \overline{x}_j)(x_{ik} - \overline{x}_k).$$

The covariance measures the degree of relation between the variables. The *correlation* between variables x_j and x_k is

$$r_{jk} = \frac{s_{jk}}{\sqrt{s_j^2 s_k^2}}.$$

Covariances are often summarized as a *covariance matrix*

$$\mathbf{S} = \begin{bmatrix} s_1^2 & \cdots & & s_{1p} \\ s_{21} & s_2^2 & & \\ & & \ddots & \\ s_{p1} & \cdots & & s_p^2 \end{bmatrix}$$

and the correlations can be summarized as a *correlation matrix*:

$$\mathbf{R} = \begin{bmatrix} 1 & \cdots & & r_{1p} \\ r_{21} & 1 & & \\ & & \ddots & \\ r_{p1} & \cdots & & 1 \end{bmatrix}$$

9 Introduction to multivariate methods

Covariance matrices and correlation matrices are symmetric.

9.2 Principal component analysis

9.2.1 Introduction

Principal components analysis (PCA) is used to summarize data about a large number of variables into a smaller number of derived variables called principal components.

Example 9.2 *In Table 9.3 you can see the national records (1984) for men in a number of running events. We might want to summarize these data into some kind of "index" for each country: which countries are "good" at running?*

This could be done in several ways.

1. Add the times for all events.
2. Compute some kind of weighted sum of the times for each country.

In the second case: how should the weights be selected? One approach is as follows.

9.2.2 Definition of principal components

Denote the different variables (running events) with $x_1, x_2, ..., x_p$. Calculate a weighted sum using the weights $\ell_{11}, \ell_{21}, \ldots \ell_{p1}$ as:

$$y_1 = \ell_{11}x_1 + \ell_{21}x_2 + \ldots \ell_{p1}x_p$$

We select the weights $\ell_{11}, \ell_{21}, \ldots \ell_{p1}$ so that the sums of their squares is 1 while at the same time the variance of y_1 is as large as possible. This is called the first principal component.

We can define the second principal component as another weighted sum of the variables:

$$y_2 = \ell_{12}x_1 + \ell_{22}x_2 + \ldots \ell_{p2}x_p$$

that has the following properties:

1. The variance of y_2 is as large as possible, under the constraint that
2. y_1 and y_2 are uncorrelated, and

Table 9.3: National running records in 1984

Country	M100	M200	M400	M800	M1500	M5000	M10000	Marath
Argentina	10.39	20.81	46.84	1.81	3.70	14.04	29.36	137.72
Australia	10.31	20.06	44.84	1.74	3.57	13.28	27.66	128.90
Austria	10.44	20.81	46.82	1.79	3.60	13.26	27.72	135.90
Belgium	10.34	20.68	45.04	1.73	3.60	13.22	27.45	129.95
Bermuda	10.28	20.58	45.91	1.80	3.75	14.68	30.55	146.62
Brazil	10.22	20.43	45.21	1.73	3.66	13.62	28.62	133.13
Burma	10.64	21.52	48.30	1.80	3.85	14.45	30.28	139.95
Canada	10.17	20.22	45.68	1.76	3.63	13.55	28.09	130.15
Chile	10.34	20.80	46.20	1.79	3.71	13.61	29.30	134.03
China	10.51	21.04	47.30	1.81	3.73	13.90	29.13	133.53
Colombia	10.43	21.05	46.10	1.82	3.74	13.49	27.88	131.35
Cook_Is	12.18	23.20	52.94	2.02	4.24	16.70	35.38	164.70
Costa_Rica	10.94	21.90	48.66	1.87	3.84	14.03	28.81	136.58
Czech	10.35	20.65	45.64	1.76	3.58	13.42	28.19	134.32
Denmark	10.56	20.52	45.89	1.78	3.61	13.50	28.11	130.78
Dom_Rep	10.14	20.65	46.80	1.82	3.82	14.91	31.45	154.12
Finland	10.43	20.69	45.49	1.74	3.61	13.27	27.52	130.87
France	10.11	20.38	45.28	1.73	3.57	13.34	27.97	132.30
GDR	10.12	20.33	44.87	1.73	3.56	13.17	27.42	129.92
FRG	10.16	20.37	44.50	1.73	3.53	13.21	27.61	132.23
GB	10.11	20.21	44.93	1.70	3.51	13.01	27.51	129.13
Greece	10.22	20.71	46.56	1.78	3.64	14.59	28.45	134.60
Guatemala	10.98	21.82	48.40	1.89	3.80	14.16	30.11	139.33
Hungary	10.26	20.62	46.02	1.77	3.62	13.49	28.44	132.58
India	10.60	21.42	45.73	1.76	3.73	13.77	28.81	131.98
Indonesia	10.59	21.49	47.80	1.84	3.92	14.73	30.79	148.83
Ireland	10.61	20.96	46.30	1.79	3.56	13.32	27.81	132.35
Israel	10.71	21.00	47.80	1.77	3.72	13.66	28.93	137.55
Italy	10.01	19.72	45.26	1.73	3.60	13.23	27.52	131.08
Japan	10.34	20.81	45.86	1.79	3.64	13.41	27.72	128.63
Kenya	10.46	20.66	44.92	1.73	3.55	13.10	27.80	129.75
Korea	10.34	20.89	46.90	1.79	3.77	13.96	29.23	136.25
P_Korea	10.91	21.94	47.30	1.85	3.77	14.13	29.67	130.87
Luxemburg	10.35	20.77	47.40	1.82	3.67	13.64	29.08	141.27
Malaysia	10.40	20.92	46.30	1.82	3.80	14.64	31.01	154.10
Mauritius	11.19	22.45	47.70	1.88	3.83	15.06	31.77	152.23
Mexico	10.42	21.30	46.10	1.80	3.65	13.46	27.95	129.20
Netherlands	10.52	20.95	45.10	1.74	3.62	13.36	27.61	129.02
NZ	10.51	20.88	46.10	1.74	3.54	13.21	27.70	128.98
Norway	10.55	21.16	46.71	1.76	3.62	13.34	27.69	131.48
Png	10.96	21.78	47.90	1.90	4.01	14.72	31.36	148.22
Philippines	10.78	21.64	46.24	1.81	3.83	14.74	30.64	145.27
Poland	10.16	20.24	45.36	1.76	3.60	13.29	27.89	131.58
Portugal	10.53	21.17	46.70	1.79	3.62	13.13	27.38	128.65
Rumania	10.41	20.98	45.87	1.76	3.64	13.25	27.67	132.50
Singapore	10.38	21.28	47.40	1.88	3.89	15.11	31.32	157.77
Spain	10.42	20.77	45.98	1.76	3.55	13.31	27.73	131.57
Sweden	10.25	20.61	45.63	1.77	3.61	13.29	27.94	130.63
Switzerlan	10.37	20.45	45.78	1.78	3.55	13.22	27.91	131.20
Taipei	10.59	21.29	46.80	1.79	3.77	14.07	30.07	139.27
Thailand	10.39	21.09	47.91	1.83	3.84	15.23	32.56	149.90
Turkey	10.71	21.43	47.60	1.79	3.67	13.56	28.58	131.50
USA	9.93	19.75	43.86	1.73	3.53	13.20	27.43	128.22
USSR	10.07	20.00	44.60	1.75	3.59	13.20	27.53	130.55
W_Samoa	10.82	21.86	49.00	2.02	4.24	16.28	34.71	161.83

9 Introduction to multivariate methods

3. The sum of squares of the weights is 1.

We could go on and define the third, fourth etc. principal component in a similar way.

It turns out that:

1. The weights ℓ_{ij} for the first, second etc. principal component are equal to the elements of the first, second etc. *eigenvector* of the covariance matrix **S**.

2. The corresponding *eigenvalues* are equal to the variance of the respective principal component.

Eigenvalues and eigenvectors are mathematical concepts in matrix algebra; they can be computed from square symmetric matrices like correlation or covariance matrices.

9.2.3 Explained variance

If you omit the last few principal components, the analysis has resulted in a "simplification" of the data. Instead of describing p variables you can look at k components, where often k is much smaller than p, hopefully without losing too much information. A measure of the success of the analysis is the percentage of "explained variance". The explained variance after the first k principal components is

$$\frac{\sum_{i=1}^{k} \lambda_i}{tr(\mathbf{S})} = \frac{(\text{Sum of the } k \text{ first eigenvalues of } \mathbf{S})}{(\text{Sum of all diagonal elements of } \mathbf{S})}.$$

If the variables you study have different scales it is customary to base the analysis on standardized variables (i.e. all variables are transformed to mean zero and variance 1). Otherwise variables that have "large" values will get too much weight in the analysis. To use standardized variables is the same as using the correlation matrix (instead of the covariance matrix) in the analysis.

9.2.4 Analysis of the example data

We return to the data from Example 9.2. Selected parts of a SAS printout of an analysis of the data on running records is as follows. The analysis was made on standardized data, i.e. the correlation matrix was used. First we get simple descriptive statistics:

Principal component analysis

```
                    Principal Component Analysis
        55 Observations
         8 Variables
                              Simple Statistics
                       M100            M200            M400            M800
        Mean       10.47109091     20.94018182     46.43872727      1.793272727
        StD         0.35142921      0.64478676      1.45701757      0.063684830
                      M1500           M5000          M10000          MARATH
        Mean        3.698181818    13.84581818     28.99672727    136.6349091
        StD         0.155909410     0.80116048      1.80168548      9.2173594
```

The correlation matrix is as follows:

```
Correlation Matrix
          M100    M200    M400    M800    M1500   M5000   M10000  MARATH
M100    1.0000  0.9225  0.8411  0.7560  0.7002  0.6195  0.6345  0.5199
M200    0.9225  1.0000  0.8507  0.8065  0.7751  0.6955  0.6970  0.5952
M400    0.8411  0.8507  1.0000  0.8702  0.8353  0.7786  0.7853  0.7044
M800    0.7560  0.8065  0.8702  1.0000  0.9180  0.8636  0.8677  0.8063
M1500   0.7002  0.7751  0.8353  0.9180  1.0000  0.9281  0.9337  0.8655
M5000   0.6195  0.6955  0.7786  0.8636  0.9281  1.0000  0.9739  0.9323
M10000  0.6345  0.6970  0.7853  0.8677  0.9337  0.9739  1.0000  0.9432
MARATH  0.5199  0.5952  0.7044  0.8063  0.8655  0.9323  0.9432  1.0000
```

We can note that the correlation between e.g. M100 and M200 is high. This means that countries that have good 100-meter runners also have good 200-meter runners. The eigenvalues of the correlation matrix are as follows:

```
              Eigenvalues of the Correlation Matrix
          Eigenvalue     Difference     Proportion     Cumulative
PRIN1       6.62100        5.74368        0.827625       0.82762
PRIN2       0.87732        0.71692        0.109665       0.93729
PRIN3       0.16040        0.03635        0.020050       0.95734
PRIN4       0.12406        0.04412        0.015507       0.97285
PRIN5       0.07993        0.01202        0.009992       0.98284
PRIN6       0.06791        0.02163        0.008489       0.99133
PRIN7       0.04628        0.02318        0.005785       0.99711
PRIN8       0.02310           .           0.002888       1.00000
```

The column "Cumulative" shows that the first principal component explains 82.8% of the variation. This is unusually high. The second component explains an additional 11%, i.e. the first two principal components together explain 93.7% of the variation.

The components can be explained by studying the loadings ℓ_{ij}. These are elements of the eigenvectors:

```
             Principal Component Analysis
                      Eigenvectors
              PRIN1        PRIN2        PRIN3        PRIN4
M100        0.317726     0.565906     0.336515     0.130252
M200        0.337017     0.461601     0.355609    -0.265436
M400        0.355573     0.249056    -0.557502     0.653788
M800        0.368657     0.012649    -0.534732    -0.474954
M1500       0.372820    -0.139714    -0.157350    -0.404230
M5000       0.364401    -0.312263     0.184737     0.028275
M10000      0.366682    -0.305929     0.196714     0.079247
MARATH      0.341893    -0.440302     0.256845     0.298457
              PRIN5        PRIN6        PRIN7        PRIN8
M100        0.284963    -0.578629     0.131288     0.123856
M200       -0.176429     0.651123    -0.104092    -0.102506
M400       -0.227499     0.146128    -0.003651    -0.012960
M800        0.542297     0.000754    -0.237552    -0.043531
M1500      -0.482629    -0.184550     0.605307     0.141204
M5000      -0.251774    -0.134923    -0.598256     0.543390
M10000     -0.117310    -0.232198    -0.171594    -0.793633
MARATH      0.481398     0.337187     0.402161     0.164393
```

9 Introduction to multivariate methods

In the first principal component (PRIN1), the variables have nearly equal weight. This component can be interpreted as a general "running ability" for the different countries. Since a short running time means a good runner, countries with low values on this component are countries that have good runners. The second component seems to measure differences between "long-distance" and "short-distance" running. Countries that have a high score on this component are countries that have runners that are specialized.

The score of each country on each of the components can be calculated. These are the scores on the first component:

OBS	COUNTRY	PRIN1							
1	USA	-3.4338	19	Spain	-1.4834	38	China	0.4071
2	GB	-3.0274	20	Czech	-1.3750	39	Israel	0.4328
3	Italy	-2.7300	21	Japan	-1.2409	40	Bermuda	0.7387
4	USSR	-2.6299	22	Hungary	-1.2076	41	Taipei	0.9494
5	GDR	-2.5932	23	Rumania	-1.1993	42	P_Korea	1.6822
6	FRG	-2.5557	24	Denmark	-1.1158	43	Malaysia	1.7083
7	Australi	-2.4271	25	Portugal	-0.9196	44	Dom_Rep	1.7151
8	France	-2.1746	26	Ireland	-0.8869	45	Burma	1.9709
9	Kenya	-2.0859	27	Norway	-0.8143	46	Philippi	2.0701
10	Belgium	-2.0443	28	Austria	-0.8104	47	Costa_Ri	2.2947
11	Poland	-2.0034	29	Mexico	-0.6814	48	Guatemal	2.6715
12	Canada	-1.7491	30	Colombia	-0.3928	49	Indonesi	2.7475
13	Finland	-1.6950	31	Chile	-0.3829	50	Thailand	2.7626
14	Switzerl	-1.6469	32	Greece	-0.3820	51	Singapor	3.1224
15	Sweden	-1.6060	33	India	-0.1673	52	Png	3.9094
16	NZ	-1.6026	34	Korea	0.2057	53	Mauritiu	4.2593
17	Brazil	-1.5605	35	Luxembur	0.2187	54	W_Samoa	7.2339
18	Netherla	-1.5583	36	Argentin	0.2603	55	Cook_Is	10.5592
			37	Turkey	0.2638			

It is often useful to plot the first principal components against each other. A Minitab plot of this type (but where the sign of component 1 has been switched) is given in Figure 9.1.

Another useful type of plot is to plot the loadings for the components against each other. Such a loading plot is given in Figure 9.2.

9.2.5 Number of components to retain

PCA is mainly used as a method to find structure in data. The number of components that are retained in the analysis has some influence on the interpretation. There are no formal tests of the number of components. However, a few commonly used rules of thumb have been established:

1. One approach is to include components until a specified amount of explained variation has been achieved, for example 90%.

2. The sum of all eigenvalues (i.e. the total variation) is equal to p, the number of variables, if the analysis is based on a correlation matrix. Thus, the average size of the eigenvalues is 1. It would then be reasonable to keep the components that correspond to eigenvalues larger than one.

Principal component analysis

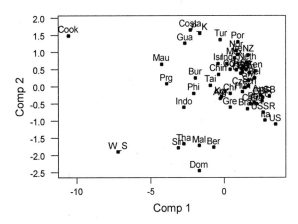

Figure 9.1: Plot of component 1 vs component 2.

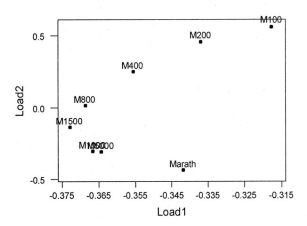

Figure 9.2: Loading plot for the first two components.

9 Introduction to multivariate methods

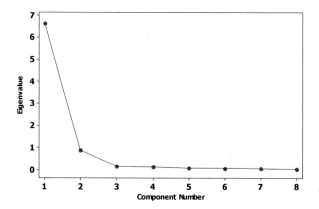

Figure 9.3: Scree plot for the data on national records in running events.

3. A scree plot (see Figure 9.3) can be used to help determine the number of components. This is a plot of the eigenvalues aginst component order. You would stop extracting more components when the scree plot has reached a "scree".

For the data on national records in running events, the first component accounts for 82.7%, and the first two components for 93.7% of the variance. There is no reason to include more than two components. The two first eigenvalues are 6.62 and 0.88; it may even be possible to stop after one component. On the other hand, the scree plot is rather "flat" from component 3 and onwards. This suggests that two components are appropriate.

9.3 Factor analysis

Factor analysis means that you set up a *model* for the way data were generated. The model contains so-called *Latent variables* or *factors*.

Example: In decathlon, the following events are included:
100 m, long jump, shot put, high jump, 400 m, 110 m hurdles, discus, pole vault, javelin and 1500 m.

Suppose that you formulate a hypothesis that only *one* attribute F affects performance in all events, apart from random variation. A model of this type may be written

$$\begin{bmatrix} x_1 \\ x_2 \\ \vdots \\ x_p \end{bmatrix} = \begin{bmatrix} \ell_{11} \\ \ell_{21} \\ \vdots \\ \ell_{p1} \end{bmatrix} F + \begin{bmatrix} e_1 \\ e_2 \\ \vdots \\ e_p \end{bmatrix}$$

x_1, x_2 etc. are the results in the events for an individual, F is the "sporting capacity" of that individual, ℓ_{ij} are weights (so called factor loadings), and e_i are random errors, sometimes called "specific factors". If you could know the value of F for an individual, you would be able to predict his/her achievement in any event, apart from random errors. The success of the prediction would depend on the size of the random errors.

But it may not be enough with *one* factor. Maybe one factor is used for the "speed" events, and another for the "power" events. Then the model would be

$$\begin{bmatrix} x_1 \\ x_2 \\ \vdots \\ x_p \end{bmatrix} = \begin{bmatrix} \ell_{11} & \ell_{12} \\ \ell_{21} & \ell_{22} \\ \vdots & \vdots \\ \ell_{p1} & \ell_{p2} \end{bmatrix} \begin{bmatrix} F_1 \\ F_2 \end{bmatrix} + \begin{bmatrix} e_1 \\ e_2 \\ \vdots \\ e_p \end{bmatrix}$$

or in general

$$\mathbf{x} = \mathbf{LF} + \epsilon.$$

Factor analysis means to estimate the parameters in this type of model, i.e. the factor loadings and the variances of ϵ. A problem here is that if L is a solution, then LT is also a solution, where T is an orthogonal transformation matrix (i.e. $TT' = I$). Thus, there is some degree of arbitrariness in the solution: we have freedom to "rotate" it. There are criteria for this "rotation" that make the solution easy to interpret.

An example (based on 160 decathlon athletes):

```
Initial Factor Method: Iterated Principal Factor Analysis
              Prior Communality Estimates: ONE
          Preliminary Eigenvalues:   Total = 10   Average = 1
                     1          2          3          4          5
Eigenvalue        3.7866     1.5173     1.1144     0.9134     0.7201
Difference        2.2693     0.4029     0.2010     0.1933     0.1251
Proportion        0.3787     0.1517     0.1114     0.0913     0.0720
Cumulative        0.3787     0.5304     0.6418     0.7332     0.8052
                     6          7          8          9         10
Eigenvalue        0.5950     0.5267     0.3837     0.2353     0.2075
Difference        0.0683     0.1430     0.1484     0.0278
Proportion        0.0595     0.0527     0.0384     0.0235     0.0207
Cumulative        0.8647     0.9173     0.9557     0.9793     1.0000
```

9 Introduction to multivariate methods

```
Initial Factor Method: Iterated Principal Factor Analysis
           Eigenvalues of the Reduced Correlation Matrix:
             Total = 5.42196805  Average = 0.54219681
                    1          2          3          4          5
Eigenvalue      3.3321     1.2708     0.8250     0.3793     0.0681
Difference      2.0613     0.4457     0.4457     0.3112     0.0616
Proportion      0.6146     0.2344     0.1522     0.0700     0.0126
Cumulative      0.6146     0.8489     1.0011     1.0711     1.0836
                    6          7          8          9         10
Eigenvalue      0.0066    -0.0514    -0.0730    -0.1599    -0.1758
Difference      0.0580     0.0216     0.0869     0.0159
Proportion      0.0012    -0.0095    -0.0135    -0.0295    -0.0324
Cumulative      1.0848     1.0754     1.0619     1.0324     1.0000
                        Factor Pattern
                        FACTOR1    FACTOR2    FACTOR3
             M100       0.67263    0.06800   -0.44268
             LONG       0.75697    0.10289   -0.21369
             SHOT       0.72160   -0.44916    0.33759
             HIGH       0.59521    0.09516    0.01963
             M400       0.58657    0.42240   -0.19656
             M110H      0.61145   -0.00770   -0.11784
             DISCUS     0.59095   -0.33978    0.31287
             POLE       0.45183    0.08738    0.09305
             JAVE       0.36930   -0.18500    0.23388
             M1500      0.15491    0.84199    0.50526
                Variance explained by each factor
                     FACTOR1    FACTOR2    FACTOR3
                    3.332113   1.270763   0.825039

           Final Communality Estimates: Total = 5.427916
            M100       LONG       SHOT       HIGH       M400
         0.653022   0.629260   0.836427   0.363716   0.561122
           M110H     DISCUS       POLE       JAVE      M1500
         0.387812   0.562569   0.220444   0.225310   0.988234
Rotation Method: Varimax
                Orthogonal Transformation Matrix
                         1          2          3
                 1    0.77658    0.62010    0.11134
                 2    0.29545   -0.51455    0.80495
                 3   -0.55644    0.59222    0.58280

                     Rotated Factor Pattern
                        FACTOR1    FACTOR2    FACTOR3
             M100       0.78877    0.11995   -0.12836
             LONG       0.73716    0.28991    0.04256
             SHOT       0.23983    0.87851   -0.08447
             HIGH       0.47942    0.33175    0.15431
             M400       0.68969    0.02998    0.29077
             M110H      0.53813    0.31334   -0.00680
             DISCUS     0.18444    0.72657   -0.02537
             POLE       0.32493    0.29032    0.17487
             JAVE       0.10199    0.46270    0.02851
             M1500      0.08792   -0.03796    0.98948
```

The results suggest that the performance of the athletes can be explained by three factors. The first factor has large loadings (0.4 or larger) for the running events (except 1500 m). This may be a "quickness" factor. The second factor has large loadings for shot put, discus and javelin. This has to do with strength. The last factor has a large loading only for 1500 m running, and to some extent to 400 m. This may be the endurance factor.

9.4 Cluster analysis

Cluster analysis is a group of statistical methods used to form groups, clusters, of the observations in a multivariate dataset. Observations that are "similar" are placed in the same cluster.

When selecting method for the analysis, the following items warrant consideration:

- How do we measure "similarity" between observations?

- Do we calculate "similarity" from standardized or from unstandardized data?

- How do we decide that two clusters are so similar that they can be merged?

9.4.1 Measures of similarity or distance

We can illustrate different measures of similarity (or dissimilarity) in two dimensions; the generalization to more dimensions is difficult to draw but mathematically analogous. For illustration we use the two points A and B in a two-dimensional space; see Figure 9.4.

Some commonly used measures of "distance" or "dissimilarity" between points A and B are:

- Euclidian distance ("Pythagoras' theorem"): $d_{ik} = \sqrt{\sum_j (x_{ij} - x_{kj})^2}$.

- The square of the Euclidian distance.

- Pearson distance: $d_{ik} = \sqrt{\sum_j (x_{ij} - x_{kj})^2 / s_j^2}$ (The variables are standardized to the same variance).

- "Manhattan distance": $d_{ik} = \sqrt{\sum_j |x_{ij} - x_{kj}|}$. "You walk around the block".

9 Introduction to multivariate methods

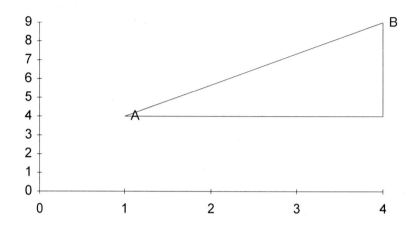

Figure 9.4: The distance between points A and B.

Sometimes you may want to measure "similarity" between variables that are not numeric. An example: We have two variables with two categories each. The data can be written as a cross table:

		Variable k		
		1	0	Total
Variable i	1	a	b	$a+b$
	0	c	d	$c+d$
Total		$a+c$	$b+d$	$p=a+b+c+d$

Using this notation, several measures of "similarity" between variables i and k may be considered:

$\frac{a+d}{p}$, the proportion of agreements

$\frac{a}{p}$, only 1-1-agreements are counted

In the cases where you want to form clusters of variables (instead of clusters of individuals), the correlation coefficient is often used as a measure of similarity between variables.

9.4.2 Measures of "distance" between clusters

You can measure the distance between clusters in several ways. Some examples:

- "Single linkage": The distance between the closest members in the

Table 9.4: Properties of diffferent brands of breakfast cereals.

Brand	Protein	Carbo	Fat	Calories	Vitamin A
Life	6	19	1	110	0
Grape nuts	3	23	0	100	25
Super sugar crisp	2	26	0	110	25
Special K	6	21	0	110	25
Rice crispies	2	25	0	110	25
Raisin bran	3	28	1	120	25
Product 19	2	24	0	110	100
Wheaties	3	23	1	110	25
Total	3	23	1	110	100
Puffed rice	1	13	0	50	0
Sugar corn pops	1	26	0	110	25
Sugar smacks	2	25	0	110	25

two clusters.

- "Average linkage": The average distance calculated from all possible pairs of members in the two clusters.

- "Complete linkage": The distance between those elements in the two clusters that are furthest apart.

- Centroid: The distance between the midpoints (average values) of the clusters.

- Median: The median distance calculated from all possible pairs of members in the two clusters.

- Ward's linkage: The sum of squared distances from the points to the cluster midpoints.

Example 9.3 *The data in Table 9.4 show the contents of some brands of breakfast cereals. Which brands are most similar?*

As an example of a cluster analysis, we analyzed these data using Minitab. The options chosen were to standardize all variables, to use Euclidean distances, and to use single-linkage clustering. Parts of the output are as follows:

9 Introduction to multivariate methods

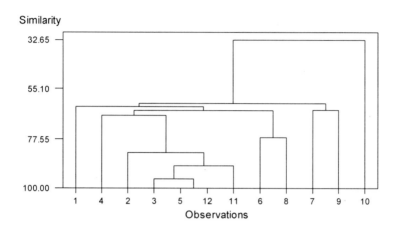

Figure 9.5: Dendrogram of the breakfast cereal data.

```
Hierarchical Cluster Analysis of Observations

Standardized Variables, Euclidean Distance, Single Linkage

Amalgamation Steps

Step Number of Similarity Distance Clusters   New     Number of obs.
     clusters   level      level    joined    cluster in new cluster
  1     11      100.00     0.000    5   12    5           2
  2     10       95.78     0.253    3    5    3           3
  3      9       89.85     0.609    3   11    3           4
  4      8       83.83     0.970    2    3    2           5
  5      7       76.94     1.383    6    8    6           2
  6      6       67.05     1.977    2    4    2           6
  7      5       64.88     2.107    2    6    2           8
  8      4       64.40     2.135    7    9    7           2
  9      3       62.91     2.225    1    2    1           9
 10      2       61.61     2.303    1    7    1          11
 11      1       32.65     4.040    1   10    1          12
```

The results from a cluster analysis is often presented as a dendrogram; see Figure 9.5.

9.4.3 Clustering variables

You can "invert" a cluster analysis and form clusters of variables instead of clusters of observations. Correlations are often used as a measure of "similarity".

Cluster analysis

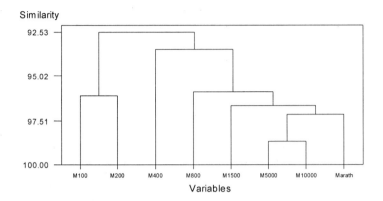

Figure 9.6: Dendrogram of the classification of different running events.

Example 9.4 *Data on national records in all track events from 100 m to marathon were given in the table on Page 217. These data can be analyzed to find similarities between the running events. A Minitab output is as follows:*

```
Correlation Coefficient Distance, Single Linkage

Amalgamation Steps

Step Number of Similarity Distance Clusters  New     Number of obs.
     clusters   level      level    joined   cluster in new cluster
  1     7       98.69      0.026    6   7     6           2
  2     6       97.16      0.057    6   8     6           3
  3     5       96.69      0.066    5   6     5           4
  4     4       96.13      0.077    1   2     1           2
  5     3       95.90      0.082    4   5     4           5
  6     2       93.51      0.130    3   4     3           6
  7     1       92.53      0.149    1   3     1           8
```

A dendrogram for these data is given in Figure 9.6.

9.4.4 Other clustering methods

The examples given above are rather "blind", i.e. they do not use any advance information that you may have on the data. Two other clustering methods are briefly summarized here.

9 Introduction to multivariate methods

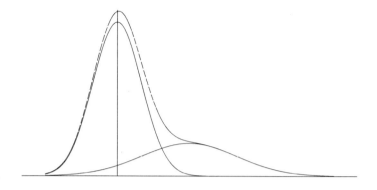

Figure 9.7: A mixture of two normal distributions.

K-means clustering

1. Divide the observations into k clusters.

2. Go through the observations one by one. Move the observation to the cluster for which the midpoint is "closest". Re-calculate mean values for the cluster that got the observation and for the cluster that lost it.

3. Repeat step 2 until no observations change cluster.

Mixtures of normal distributions

Sometimes you may suspect that a given set of data may have been obtained by mixing data from several populations. If it can be assumed that the population distributions are normal, it may be possible to subdivide the data into populations. In Figure 9.7, the idea is illustrated in one dimension.

Figure 9.7 shows two normal distributions. The one to the left has mean value 0 and standard deviation 1, and the one to the right has mean 3 and standard deviation 4. These two distributions have been mixed (30% from the left, 70% from the right distribution). The resulting mixed distribution has two peaks.

In practice you can only observe the mixed distribution. However, there are methods to estimate the parameters of the distributions involved, and their mixing ratio. The method results in statistical tests of the hypothesis that data come from 1, 2, 3.... normal distributions.

9.4.5 Some advice on choice of method

- If the clusters are well separated most methods work well.

- If you use average, centroid, median or Ward linkage, squared Euclidean (or squared Pearson) distances should be used.

- Try several measures of distance. If you obtain similar results, your conclusions are more robust.

- Single, centroid and median linkage seem to be sensitive to measurement errors, but not very sensitive to "outliers". Ward and average linkage are sensitive to outliers, but not so sensitive to measurement errors.

9 Introduction to multivariate methods

9.5 Exercises

Exercise 9.1 *The following table is a subset of a larger table. A panel of consumers were asked to rate different apple varieties with respect to juiciness, hardness, sweetness, acidity, bitterness and zestyness. The results were averaged over all members in the panel into a score for each variety. The Euclidian distances between the varieties were calculated based on these scores. The plan is to make a "single linkage" cluster analysis of the data. The matrix of Euclidian distances is as follows:*

	Splendour	Granny Smith	Top red	Celiste	Golden Delicious
Splendour	0				
Granny S	29.1	0			
Top red	13.2	22.9	0		
Celiste	32.7	45.6	29.6	0	
Golden Del.	36.9	57.1	37.9	17.9	0

A. Which varieties will first be merged into a cluster?

B. In the second step of the cluster analysis, which cluster will be formed?

C. Complete a single linkage, hierarchical cluster analysis of the data. Present the results as a table and as a dendrogram.

Chapter 10

Topics in clinical trials

10.1 Introduction to clinical trials

10.1.1 Terminology and brief historical overview

A *clinical trial* is a type of experiment where different medical treatments are compared on human subjects in a systematic way. The treatments may be different operation methods, clinical procedures, drugs, etc. Many clinical trials are performed by pharmaceutical companies as part of their research on new drugs.

The scientific ideas underlying clinical trials are of rather recent origin. Pioneering work on the use of data to support medical hypotheses was done from about 1700 onwards. Some early workers in this area include William Petty (1623-1687), John Graunt (1620-1674), Pierre Louis (1787-1872), William Farr (1807-1883), Florence Nightingale (1820-1910) and Joseph Dietl (1804-1878).

Example 10.1 *Blood letting and induced vomiting were for a long time regarded as universal therapies against a long range of diseases. The underlying idea, called humoral pathology, was that all diseases are caused by some kind of evil in one of four bodily fluids. The treatment therefore focused on removing these fluids. Dietl (1849) compared three treatments against pneumonia in an early clinical trial. The treatments were blood letting, induced vomiting, and rest (self healing). The results are reported in Table 10.1.*

The data strongly suggest that the traditional therapies are inefficient. Both blood letting and induced vomiting give a death rate of about 20%. With self healing, which includes rest and a light diet, the death rate is reduced to about 7%. The experiment was not randomized, in the modern

10 Topics in clinical trials

Table 10.1: Dietls data on treatment of pneumonia with different methods.

	Blood letting	Induced vomiting	Self healing
Survived	68	84	175
Died	17	22	14
Total	85	106	189
% deaths	20.0%	20.7%	7.4%

sense, but the data are still rather convincing, and the experiment started the decline of humoral pathology.

Modern clinical trials are *randomized* such that the decision on which patient will get which treatment is left to chance. They are often *double blind*, which means that neither the patient nor the doctor knows which treatment the patient received. Finally, if ethically defendable, they may include a *control* group, i.e. a group of patients who get a non-potent drug or treatment (*placebo*).

Sir Austin Bradford Hill (1897-1991), was involved in several early clinical trials that were based on the randomization ideas of Sir R. A. Fisher (1890-1962). According to Pocock (1983), the first clinical trial with a randomized control group was a trial of streptomycin for treatment of pulmonary tubercelosis (Medical Research Council, 1948). An early example of the use of placebo in a trial was a clinical trial to study the effect of antihistamines on common colds (Medical research council, 1950). In this study, no significant differences were found between the placebo group and the antihistamine group.

10.1.2 Methods of investigation in clinical research

Chance observation

Some medical discoveries have been made by chance: a researcher has discovered an unusual pattern and has drawn conclusions from this. Some examples:

Example 10.2 *Discovery of Digitalis: Withering (1775) discovered the relation between Digitalis and heart problems when the drug was used for other purposes.*

Example 10.3 *Discovery of Penicillin: Fleming observed an unusual contamination from the air of some plates lying in his lab.*

Introduction to clinical trials

Example 10.4 *Aspirin - heart attacks: L. L. Craven observed gum bleeding in children chewing asprin gum, concluding that aspirin might be an anticuagulant for use in heart attack patients.*

Case histories

Case histories of "interesting" individual cases are often described in medical journals. Series of case histories may be described e.g. in tables. In some cases, conclusions that can be generalized to other patients are drawn from case histories.

Uncontrolled trials

New treatments are sometimes tried, with or without control groups, on "real" patients. However, ethical considerations makes it difficult to get permission to conduct such "ad hoc"-studies nowadays.

Example 10.5 *In 1605, the British East India Company sent three ships to India. In those days, scurvy was a major cause of death during long ocean passages. It was beleived that this disease was caused by lack of fresh food. On one of the ships, lemon juice was distributed to the sailors as soon as they fell ill. On the other two ships, no such treatment was distributed. The results are summarized in Table 10.2.*

Table 10.2: Number of deaths and survivors on the British East India Company expedition to India in 1605.

Number of	Lemon juice	No treatment
Deaths	0	105
Survivors	202	117
Total	202	222

Although not all deaths were caused by scurvy, and although the sailors were not randomly distributed on treatments, the data seem to suggest that lemon juice treatment may be effective. This conclusion, however, was not made until a hundred years later when all British sailors got lemon juice as part of their diet (and were, therefore, called "Limeys" by sailors in other navies).

Planned observation with comparison groups ("observational studies")

In observational studies, patients with possibly different treatments are followed. The treatments are not decided as part of the study but are

10   Topics in clinical trials

the treatments the patients happened to get. Thus, in an observational study the data for each patient are recorded, but no intervention is done.

Cross-sectional studies   A cross-sectional study is an observational study made at a single point in time. The purpose may be to estimate the distribution of some quantity, such as the distribution of a disease, in some population. In cross-sectional studies, this is often achieved by selecting a random sample from the population at a certain time point. In cases where the development over time is studied, new samples are taken at every time point.

Cross-sectional studies are relatively cheap and can give many measurements. In some cases, data may naturally exist in files or registers. Cross-sectional studies are less useful for rare conditions, since the chance of finding a case during the study period is low.

Case-control studies   In a case-control study, data are obtained for a group of patients with a specific disease. For each patient, one or more controls are selected that match the patient in terms of important background variables (age, sex etc.). The aim is to get a control group that is as similar as possible to the patient group. Case-control studies are often used in epidemiology.

Case-control studies are useful for study of rare diseases. They are relatively quick, simple and cheap. In a case-control study, the relation between the risk of disease and several background variables can be studied.

Some disadvantages are that case-control studies rely on recall or records of the past. Validation of the data, as well as control of extraneous variables, may be difficult. The crucial point in a case-control study is the proper selection of a comparison group. Since no randomization is done, a case-control study can establish a relation between background variables and disease, but it is debatable whether causal inference can be made.

Example 10.6 *In a classical study, Doll and Hill (1954) studied the smoking habits for 649 British doctors who had developed lung cancer, and a matched sample of 649 doctors without lung cancer. The data for the male doctors are summarized in Table 10.3. The so called* relative risk $(\widehat{RR})$ *indicates that heavy smokers have a higher risk of developing lung cancer.*

Cohort studies   In *cohort studies*, a group of individuals is followed over time. During the observation period occurrences of disease are

Introduction to clinical trials

Table 10.3: Smoking habits for male British doctors with and without lung cancer. Data from Doll and Hill, 1954.

	No. of cigarettes/day			
	0	1-4	5-14	15-
Cases	2	33	250	364
Controls	27	55	293	274
$\widehat{RR}$	1.0	8.1	11.5	17.9

recorded for each individual, as well as the individual's exposure to various risk factors. In this way, measures of the risk of disease and the relation between risk and exposure can be established. Some cohort studies are *prospective*, which means that individuals are selected for participation at the start of the study and are then followed during the study. Other cohort studies are *retrospective*. In such studies individuals are selected for participation. Records of their disease history and exposure history are then used to estimate disease prevalence and the relation between disease and risk factors.

Some advantages of cohort studies are that we get a complete description of patient history. Multiple causes of disease can be studied, and it is possible to calculate rates in exposed and unexposed indviduals.

For rare diseases, cohort studies require large samples. A prospective cohort study may take a very long time to complete. In some cases, medical practices may change during the study which makes comparisons difficult. Good follow-up is difficult since we need to identify and locate the participating individuals maybe decades after the start of the study.

Randomized clinical trials

Randomized clinical trials have the following components:

1. They includes a *control group* (Comparative trials.)

2. Treatment is assigned to each patient by a random method after informed consent. The treatment is not known in advance either by the doctor or by the patient (*Double blind*).

3. Careful definition of patient *eligibility*.

4. Unbiased *endpoint* ascertainment - complete follow-up of all patients. The endpoint is the primary response variable(s) by which safety and efficiacy is judged.

10 Topics in clinical trials

10.1.3 Methods for clinical trials

Statistical methods   The methodology used for clinical trials has a strong statistical component. Many of the methods that are used in other application areas are also used for clinical trials: Anova, Ancova, t tests, regression analysis, etc. In addition, many statistical methods have been developed or modified to be used for clinical trials.

Clinical methods   The methods used also depend on clinical and ethical aspects. Clinical trials are performed on human beings. This brings particular problems to the design and analysis. Ethical aspects have a strong role: as soon as one treatment has been proved effective, it would be unethical to continue to give a patient some other treatment. Also, it would be unethical to make a clinical trial so small that no valid conclusions can be drawn.

Regulatory requirements   Finally, there are regulatory requirements. Before a drug may be sold it has to be approved by the authorities. Agencies like the Food and Drug Administration (FDA) in USA or the Medical Products Agency (Läkemedelsverket), in Sweden, have set up standards and requirements that must be followed. This affects the planning and analysis of clinical trials. Figure 10.1 illustrates the different views on clinical trials.

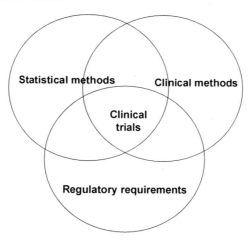

Figure 10.1: Clinical trials include statistical, clinical and regulatory components.

10.1.4 Phases of clinical trials

The development of a new drug is often described in terms of four phases: Phase I to Phase IV. These are the parts of the drug development that include human subjects. In fact, before Phase I, the drug has often undergone laboratory testing, experiments on animals, etc.

The classification into phases is to some extent overlapping. The "phase" terminology is mainly used for drugs, as opposed to operation methods and other clinical procedures. Special statistical methods have been developed for use in the different phases of the process.

Phase I trials: Pharmacology and toxicity

The absorbation, metabolism, toxicity, administration, elimination, safety, and preparation of the drug is studied. Often: Small samples, healthy young people (volonteers.) Sample sizes around 20-80.

Phase II trials: Dose-finding

Efficiacy, side effects, doses, population. Rather small samples (100-200), "real" patients. About 20% of the drugs pass this stage.

Phase III trials: Full-scale evaluation

Large samples, comparison with other treatments, randomized, efficiacy. The "real" clinical trial. Often done with several clinical centers ("multi-center trial".) Typically, it is the Phace III studies that give the basis for registration of a new drug.

Phase IV trials: Post-marketing

Post-marketing: Side effects etc.

10.1.5 Stages of a clinical trial

1. Definition of purpose; initial protocol

2. Funding

3. Detailed protocol, manual of operations, data collection forms

4. Training, pilot testing

5. Patient recruitment

6. Monitoring for quality assurance and follow-up

10 Topics in clinical trials

   7. Close-out

   8. Data analysis

10.2 Design of clinical trials

10.2.1 The purpose of a clinical trial

Medical doctors and statisticians often have different views on research. While the doctors, by training, tend to focus on the individual patient, statisticians think in terms of averages and collectives. However, these two views may often converge. The doctor should give the patient the treatment that, based on earlier experience, has the largest chance to cure the patient. Clinical trials are the tools that give that "earlier experience".

Senn (1997) defines the *effect* of a treatment as follows:

> *The effect of any treatment for a given patient is the difference between what happened to the patient as a result of giving him the treatment and what would have happened had the treatment been denied.*

This is called a *counterfactual* argument. It is a problematic argument. We can observe what happened to a patient with a certain treatment but we cannot observe what would have happened without that treatment. This is what makes it necessary in clinical studies to include a control group. If a control group is included, the effect of treatment A can be estimated, not for the individual patient but at least as an average, collective measure.

10.2.2 Variables and endpoints

In typical clinical studies, a large number of variables are collected about each patient. Some of these variables provide background information on the patient (age, sex, diagnosis etc.) while other variables are measurements taken during the study. The main outcome variable(s) in a clinical trial are called *endpoints*. A well designed study should include a rather limited number of endpoints. A clinical study on a cancer treatment may include endpoints such as length of survival, time to remission, cancer size and/or measures of side effects. Sometimes it may not be practical to measure the desired endpoint. For example, in a study on hypertension, mortality may be the desired endpoint, but mortality may be so low that enough mortality data cannot be collected during the time span of the study. Then, high blood pressure may be measured as a *surrogate endpoint* for mortality.

10.2.3 The study protocol

The *protocol* is an important document that describes how the study will be conducted. Since it includes many statistical aspects, it is crucial that a statistician is involved. In fact, the protocols used in clinical studies could serve as a model for research in other areas. In academic clinical research, the protocol is used to obtain ethical clearance for the study by the university's Ethical committee. In studies on pharmaceutical drugs, the protocol is used for communication with the country's regulatory agency.

Some items that are often included in the protocol are:

The purpose of the study.

The products being used.

Procedures that will be used on the subjects.

Eligibility of subjects.

Procedure that will be used to obtain informed consent from the subjects.

Number of subjects.

Experimental design.

How the data will be analyzed.

10.2.4 Parallel groups designs

The need for randomization

Clinical trials should be randomized, i.e. the assignment of patients or experimental subjects to the treatments should be governed by chance. There are several reasons for this. One reason is that random assignment justifies the use of statistical procedures like randomization tests. A second reason is that randomization may prevent bias that can arise, consciously or unconsciously, if the assignment of patients is left to subjective judgement.

The simplest type of randomization is the completely randomized design, in medical applications also called *parallel groups designs*. In this design, all possible combinations of patients into a treatment groups are equally likely. One way of achieving this is to use a simple lottery.

Practical randomization

The randomization can be undertaken as follows:

Example: We have 45 patients that will be randomized to three treatments. A side condition is that we want the number of patients per treatment to be equal (*balanced*). One way of achieving this kind of randomization is as follows:

1. Make a list of all patients using your favorite software (SAS, Excel, Minitab etc).

2. Use the random number generator in your software to assign a random number between 0 and 1 to each patient (rectangular distribution).

3. Sort the records by the size of the random number. The patients are now in random order.

4. Patients 1-15 on the sorted list are assigned to treatment A_1, patients 16-30 to A_2 and 31-45 to A_3.

Note that the actual names of the patients are not needed for this procedure to work: it is enough to know their sequence numbers. The patients may be numbered as they arrive at the clinic and the sequence numbers are randomized to treatments.

An alternative method is to use some random device to assign each new patient to a treatment. For example, if there are two treatments, one might flip a coin. One problem with this approach is that the treatment groups may end up to be of different size (unbalanced). It is of some advantage for the analysis if the design is balanced, i.e. if all treatment groups are of equal size.

Problems with parallel groups designs

In a clinical trial, the patient should not be aware of which treatment has been assigned to him or her; the trial should be *blinded*. Furthermore, the doctor treating the patient should also be unaware of which treatment is given to the patient; the trial should be *double blind*. Of course, if the treatment is to give an operation or not, it is not possible to seclude the treatment, but in the case of drugs, even patients receiving "no treatment" are given a non-potent pill (a *placebo*). The reasons for this secrecy is that both patients and doctors tend to act in different ways if they know that a potent medication is given compared to when a placebo is given.

In many trials with drugs, a *placebo effect* can be demonstrated. This means that even patients who get placebo will improve their results compared to some baseline.

The need for double blind experiments makes it necessary to take steps to hide the treatment allocation from the doctor. Thus, the doctor should not be the one who does the randomization. Also, the drugs for each patient should be delivered in individual unmarked envelopes.

It should be noted that randomization alone cannot eliminate bias in background variables. If treatment allocation is completely random, it is quite possible to get experimental groups that differ on important background variables such that, for example, one treatment gets more female patients than the other treatment. *Stratification* (see below) may be used to alleviate this problem. Also, even if the experiment has been randomized in a balanced way, it may be necessary to terminate data collection before the planned sample size is achieved. At that stage, the experiment may very well be unbalanced. *Blocking* (see below) is a way of overcoming this problem.

There are also ethical problems. If one treatment is thought to be inferior to the others, it may be unethical to randomize patients to this treatment.

10.2.5 Blocking

Blocking for balance

Balanced experiments are more efficient than unbalanced ones. Blocking of the observations is a useful tool to achieve (near) balance in clinical trials. The patients are then blocked into groups. The block size is often a multiple of the number of treatments. An equal number of patients is randomly allocated to each treatment within each block. Thus, for example, with two treatments and a block size of four, two patients in each block would be allocated to each treatment. This kind of blocking guarantees that the experiment will not deviate too much from a balanced experiment. On the other hand, if the doctor is aware that blocking is used, he or she may after some time be able to guess the treatment.

Blocking to reduce error variation

Blocking of the observations may be made for a second, and perhaps more important, purpose. In agricultural field experiments, plots which are located "close together" are often taken to form blocks. The experiment is then planned such that all treatments are present in each block. If the plots within a block are more "similar" than plots in a different block, comparisons between treatments are then made on plots that are as similar as possible. This will reduce the effect of possible inhomogeneity between plots, and will thus reduce the effect of error variation. The experiment would become more efficient.

This kind of blocking can sometimes be used in clinical trials as well, especially if all patients are present at the start of the study. For example, in studies on animals, the animals may be blocked by litter. However, since the patients in clinical trials often arrive consecutively, blocking in clinical studies is done mainly in order to achieve balance.

In the analysis of blocked experiments, "Block" can be included in the model as a "Class" variable (in SAS language.) Some statisticians prefer to regard "Block" as a random factor and use methods and software for Mixed models for the analysis. Normally, the block factor would not interact with other factors in the model. If it does, this is an indication that the blocks differ in some systematic way, for example that some blocks contain patients that are more severely ill than patients in other blocks. In such cases, it would have been advisable to stratify the patients.

Randomly permuted blocks

Blocked randomization ("randomly permuted blocks") can be obtained as follows. The example relates to 45 patients and three treatments. For example Excel, Minitab or SAS can be used.

1. Make a list of the patient numbers 1 to 45 in one column of the worksheet.

2. Assign a block number to each patient, in groups of three. In the example, the first seven lines of the worksheet would then be:

Patient	Block
1	1
2	1
3	1
4	2
5	2
6	2
7	3
...	

3. Assign a random number to each row of the table.

4. Sort the data lines first by block, then by the random number.

5. The first patient (in sorted order) in each block is assigned treatment A_1, the second gets A_2 and the third gets treatment A_3.

Note: Randomly permuted blocks are often used in studies where patients arrive at different times to the study. The protocol might then state e.g. that the first three patients constitute a block.

10.2.6 Stratification

In many studies, a number of prognostic factors can be recorded for the patients before they enter the study. Some examples of such prognostic factors are sex, weight, severity of the disease, pre-treatment score on some test, etc. It can often be expected that patients differing in their prognostic factors will behave in different ways during the study. If the patients are randomized to treatment without regard for the prognostic factors, there is therefore a risk that, e.g., one treatment will by chance contain more women than the other treatment.

Stratification is a method that can be used to account for such differences. One way to do this is to make the analysis separately for subgroups, e.g. one analysis for women and one for men (post-stratification.) However, stratification can be used already at the planning stage, in order to achieve (near) balance on the stratification variable. For stratification variables that are continuous (such as pre-treatment score on some test), the stratification might be done on the variable grouped into classes, such as $\begin{cases} x < 250 & \text{Group 1} \\ x \geq 250 & \text{Group 2} \end{cases}$.

Stratification is often used in the planning of clinical studies in conjunction with blocking. Suppose that the patients will be stratified by sex, and blocked by order of arrival to the study. If there are three treatments, and the block size is chosen to be three, the randomization would then be such that in each group of three men, and in each group of three women, one is randomly distributed to each of the treatments.

Stratification variables that are discrete ("class") variables can be treated in two ways. One way is to do separate analyses for each group, e.g. for each sex separately. It might be more interesting, however, to include the stratification variable in the analysis as a factor. In this way, it would be possible to detect possible interactions such as sex*treatment. Such interactions are often of clinical importance. For continuous stratification variables, methods like Analysis of covariance are often useful.

If there are many potential stratification variables, it may not be feasible to stratify by all of them at the same time. For example, a stratification by sex, age group (five classes), and disease status (three classes) would give $2 \cdot 5 \cdot 3 = 30$ strata. In a study on a rare disease with, say, a hundred patients, this would give very few, or even no, patients in some of the strata.

10.2.7 Minimization and balancing

Severeal treatment allocation schemes have been suggested in order to secure a higher degree of balance on background variables for the different treatments. Minimization is such a method. The idea is to construct some measure of imbalance of the background variables. For each new patient, this measure is recalculated assuming the patient is assigned to treatment A_1 and assuming the patient is assigned to treatment A_2. The patient is assigned to the treatment giving the smallest imbalance measure. If the imbalance is equal for all treatments the patient is randomized to one of them.

Minimization does not give the same degree of randomness as a completely randomized design or a blocked design; in fact it may happen that only the first patient is randomized. Still, Piantadosi (1997) claims that

> "... treating the data as though they arose from a completely randomized design is commonly done and probably results in few errors of inference".

Senn (1997) is more negative, stating that

> "It is also debated, however, as to whether minimisation brings any real advantages compared to randomization ... Thus, we have an example of a genuine unresolved issue".

10.2.8 Cross-over trials

Cross-over trials (in other application areas called change-over or carry-over trials) is a type of trial where each patient serves as his/her own control. This makes treatment comparisons strong, which means that a smaller sample size is needed to achieve the same statistical power. On the other hand, cross-over trials are useful mainly for experiments on diseases which are chronical, or at least long-lasting, and where the treatment cannot be expected to cure the disease, only to remedy some of its effects.

A simple cross-over trial for comparing two treatments A and B may be designed as follows. The patients at hand are blocked into groups of two. In each block, one patient is randomized to start with treatment A and the other patient starts with treatment B. After some time, the result is recorded. A wash-out period without treatment is then often be used in order to remove any effects of the treatment given in period 1. In the second period, the patients switch the treatments. The design is outlined in Table 10.4.

Table 10.4: Example of a cross-over trial with two treatments.

	Period 1	Wash-out	Period 2
Patient 1	A		B
Patient 2	B		A

In the analysis of a cross-over trial one has to account for the fact that the observations within patient are dependent. This may be handled by including "patient" as a random factor in the model. The model for analysis may also include treatment effects, period effects, carry-over effects (for example, the effect of taking treatment A before treatment B), and sequence effects. In a small trial, all these effects may not be estimable.

10.3 Reliability and validity

Doctors often regard their measurements as true and final. However, most measurements taken on patients, such as blood pressure, Hb etc. are prone to be influenced by external factors. Some such influence may be systematic, causing biased measurements, while other influences can be regarded as random. For example, when you take the blood pressure of a patient, the patient should have rested for a while and be calm and confident. An excited patient may give a higher blood pressure reading than the "normal" one.

Definition 1 *The* reliability *of a measurement is the degree to which the measurements would give the same result when applied several times on the same subjects under identical conditions.*

Definition 2 *The* validity *of a measurement is the degree to which the measurement measures what it is intended to measure.*

The reliability can be estimated by measuring e.g. the blood pressure several times on the same patients, under the same condition ("test-retest reliability"). The validity of a measurement is more difficult to assess, and will not be treated here.

10 Topics in clinical trials

Example 10.7 *Early "lie detectors" were meant to indicate whether a person was lying or not by measuring the electric resistance in the skin. The idea was that a person would sweat when lying, which would give a lower resistance. However, experiments showed that the deveice was not able to detect lies: it had low validity.*

10.3.1 A measurement model

Let us denote a measurement on one individual for some variable with y. If we could have measured the same individual again, it is probable that some different value of y would be obtained. If we could repeat the measurement process an infinite number of times under the same conditions, we could use the mean value T of these measurements as the "true" value. A simple model for a single measurement of y can be written as

$$y = T + e$$

where y is the measured value, T is the true value and e is a residual indicating the difference between y and T. If we take a population of individuals, the true scores T will vary around some mean value μ with a variance σ_T^2. For any single individual, the observed scores y will vary around T with a variance σ_e^2. Thus, the variance of y may be decomposed as

$$\begin{aligned} Var(y) &= \sigma_y^2 = Var(T+e) \\ &= Var(T) + Var(e) = \sigma_T^2 + \sigma_e^2. \end{aligned}$$

This is valid if it can be assumed that the random variables T and e are uncorrelated.

The reliability of some measurement (also called the intraclass correlation) is defined as

$$R = \frac{\sigma_T^2}{\sigma_T^2 + \sigma_e^2}.$$

If the error variance σ_e^2 is small, R will be close to one while a large value of σ_e^2 will cause the reliability coefficient to be smaller than one. If $R < 1$, the data are saud to be *attenuated*.

10.3.2 Effects of measurement errors

Effects on sample size

If the data contain measurement errors, the sample sizes required to demonstrate a certain effect is increased. It may be shown that the sample size required for a double sided t test at level α to have a power of $1-\beta$ to detect a difference between two treatments of size Δ is

$$n = \frac{2\sigma_T^2 \left(z_{1-\alpha/2} + z_{1-\beta}\right)^2}{\Delta^2}.$$

where $z_{1-\alpha/2}$ and $z_{1-\beta}$ are the limits for z which correspond to the required significance level and power, respectively.

If the reliability $R < 1$, the variance σ_T^2 is replaced by $\sigma_T^2 + \sigma_e^2$ so we need a sample size of

$$n^* = \frac{2\left(\sigma_T^2 + \sigma_e^2\right)\left(z_{1-\alpha/2} + z_{1-\beta}\right)^2}{\Delta^2} = \frac{n}{R}$$

where R is the reliability. Since R is always smaller than or equal to 1, the required sample size n^* is always larger than or equal to n.

Example 10.8 *Two treatments will be compared using a t test. It is assumed that the standard deviation of y is around $\sigma_y = 6$. The reliability of y is about 0.9. How large sample is needed if a double sided test at the 5% level should have a power of at least 0.9?*

The required sample size, without correction for attenuation, is

$$n \approx \frac{2\sigma_y^2 \left(z_{1-\alpha/2} + z_{1-\beta}\right)^2}{\Delta^2}$$

where in our example, $\Delta = 5$, $z_{1-\alpha/2} = z_{0.975} = 1.96$ and $z_{1-\beta} = z_{0.90} = 1.282$. This gives

$$n \approx \frac{2 \cdot 6_y^2 \left(1.96 + 1.282\right)^2}{5^2} = 30.27.$$

To correct for attenuation we calculate

$$n^* = \frac{n}{R} = \frac{30.27}{0.90} = 33.6.$$

We would need a sample size of 34 units.

Effects on correlations

Suppose that we have two variables x and y that both contain measurement errors. Measurement models for the two variables are

$$x = T + e$$
$$y = U + f.$$

Suppose that we are interested in the correlation between T and U. It can be shown that

$$\rho_{xy} = \rho_{TU}\sqrt{R_x R_y}$$

where R_x and R_y are the reliabilities of x and y, respectively. For data with a reliability less than 1, this is always smaller than ρ_{TU}. The correlation is said to be attenuated by measurement errors.

10.3.3 Estimating the reliability

To estimate the reliability we need data that can be used to estimate the variance components σ_T^2 and σ_e^2. A simple example of such data is the simple replication study, where the same subjects have been measured several times. If we assume that the data are balanced, the data would have the following structure:

Subject	n	$\bar{y}$	s^2
1	n	$\bar{y}_1$	s_1^2
2	n	$\bar{y}_2$	s_2^2
$\vdots$			
a	n	$\bar{y}_a$	s_a^2
Total	$N = na$	$\bar{y}$	s^2

A simple one-way ANOVA on this type of data would give the following ANOVA table:

Source	df	SS	MS	$E(MS)$
Subjects	$a - 1$	$n\sum(\bar{y}_i - \bar{y})^2$	MS_A	$\sigma_e^2 + n\sigma_T^2$
Error	$N - a$	$(n-1)\sum s_i^2$	MS_e	σ_e^2
Total	$N - 1$	$\sum(y_{ij} - \bar{y})^2$		

Based on the $E(MS)$-column, we can obtain estimators of the two variance components as

$$\widehat{\sigma}_e^2 = MS_e, \text{ and}$$
$$\widehat{\sigma}_T^2 = \frac{MS_A - MS_e}{n}.$$

This gives, as estimator of the reliability coefficient,

$$\widehat{R} = \frac{\widehat{\sigma}_T^2}{\widehat{\sigma}_T^2 + \widehat{\sigma}_e^2} = \frac{MS_A - MS_e}{MS_A + (n-1)MS_e}.$$

If the data are unbalanced, and in more complicated experimental designs, the variance components can be estimated using methods for so called Mixed models. An example of software that can be used for mixed models is the Mixed procedure of the SAS package, or similar R or SPSS tools.

Example 10.9 *Each of eight patients were measured twice for a certain test. Results:*

	\multicolumn{8}{c}{Patient}							
	1	2	3	4	5	6	7	8
	10	14	8	14	13	10	8	16
	12	15	6	18	11	12	9	18
Σ	22	29	14	32	24	22	17	34	194

To estimate the reliability we need to compile an ANOVA table: $SS_A = 5050/2 - 2352.25 = 172.75$ *with 7 df.* $SS_T = 2544 - 2352.25 = 191.75$ *with 15 df.*

Source	df	SS	MS	E(MS)
Patient	7	172.75	24.68	$\sigma_e^2 + 2\sigma_T^2$
Residual	8	19.00	2.375	σ_e^2
Total	15	191.75		

$\widehat{\sigma}_e^2 = 2.375$ is an estimate of the error variance. $\widehat{\sigma}_T^2 = (24.68 - 2.375)/2 = 11.153$ is an estimate of the variance component for patients. The reliability can be estimated as

$$\widehat{R} = \frac{11.153}{2.375 + 11.153} = 0.82.$$

ns in clinical trials

10.4 Early stopping

Clinical trials may sometimes be stopped earlier than planned. Some reasons for this are:

- Adverse effects are encountered
- The benefit from the treatment is greater than expected; it would be unethical to continue the trial
- It seems improbable that the completed trial will attain a significant effect
- Data quality problems

Suppose that some response variable is measured in a clinical trial with two groups (E and C). Half way through the trial it is decided that H$_0$ will be tested on the available data. How does that affect the level of the test?

It can be shown that, if the tests are performed on equal parts of the data the total level of the tests, taken together, is:

1 test	level 0.05
2 tests	level 0.08
5 tests	level 0.14
10 tests	level 0.20

Thus, repeated testing during the trial changes the significance level of the testing procedure. Technical steps might be needed to make sure that sound decisions are taken.

10.4.1 Methods to make conclusions during the trial

Sequential methods

In a sequential design, a "test" is performed after each patient has been reviewed. Does the new observation favour E or C? This is evaluated using the likelihood function

$$L = \prod_1^n p^k(1-p)^{n-k}.$$

The evidence for E against C can be evaluated through the likelihood ratio

$$R = \frac{L_1}{L_2} = \left(\frac{p_1}{p_2}\right)^k \left(\frac{1-p_1}{1-p_2}\right)^{n-k}$$

where p_1 and p_2 are clinically relevant values of the response probability. If R is "large", $(>R_U)$, evidence favours group 1; if R is "small", $(<R_L)$ evidence favours group 2. The stopping criteria can be expressed as proportions (k/n) rather than numbers (k). It holds that

$$\log R = k \log \frac{p_1}{p_2} + (n-k) \log \left(\frac{1-p_1}{1-p_2}\right)$$

which gives

$$k = \frac{\log R - n \log \left(\frac{1-p_1}{1-p_2}\right)}{\log \Theta}$$

or equivalently

$$\frac{k}{n} = \frac{\frac{\log R}{n} - \log \left(\frac{1-p_1}{1-p_2}\right)}{\log \Theta}$$

where $\Theta = \frac{p_1/(1-p_1)}{p_2/(1-p_2)}$ is the odds ratio. Limits for R may be taken, from experience, as $R_U = 32$ (strong evidence), or $R_U = 8$ (weak evidence). Alternatively, limits can be chosen from

$$R_U = (1-\alpha)/\beta, \text{ and}$$
$$R_L = \alpha/(1-\beta)$$

where α and $(1-\beta)$ are the required size and power of the test (Wald, 1947). As an example, suppose that $p_1 = 0.15$ and $p_2 = 0.30$. We want to use $\alpha = 0.05$ and $\beta = 0.20$ which gives $R_U = 0.95/0.20 = 4.75$ and $R_L = 0.05/0.80 = 0.0625$. A graph of the different regions for decisions is given in Figure 10.2.

Similar methods can be constructed for event rates, and for continuous response.

Group sequential methods

Method: Divide the patients into k equal-sized groups with $2n$ participants in each. Randomize treatments within each group (n patients to each treatment, total sample size: $2nk$). Test the hypothesis when each group is completed (based on $2n$, $4n$, ... patients). The corresponding test statistics z_i are compared to stopping boundaries $\pm z'_k$ which are determined such that the overall significance level for the k tests is, for example, 0.05.

Determination of these limits can be done in different ways. Pocock uses fixed limits in each step (for example, 2.413 for $k = 5$, $\alpha = 0.05$).

10 Topics in clinical trials

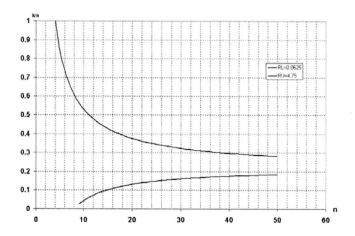

Figure 10.2: A sequential test with $p_1 = 0.15$, $p_2 = 0.30$, $R_L = 0.0625$ and $R_U = 4.75$.

Other authors have noted that the use of different limits in the different steps will produce a "final" test that has a level more similar to the nominal level 0.05. A set of limits for some such procedures is given in Figure 10.3 and Table 10.5 below.

10.4.2 Alpha spending functions

A problem with group sequential methods is that k has to be specified in advance; there is no room for new ideas during the trial. Also, the design must be balanced. As a remedy, Lan and De Mets have proposed the Alpha spending function approach.

Suppose that R interim analyses are planned. The total amount of information is I. Let i_j be the information available at the j:th analysis, and let $\tau_j = i_j/I$ be the information fraction. Commonly, for studies comparing mean values, $\tau = n/N$ where n is the interim accrual and N is the total sample size.

The alpha spending function $\alpha(t)$ is a smooth function such that $\alpha(0) = 0$ and $\alpha(1) = \alpha$, the final type 1 error rate. Different choices of alpha spending functions have been proposed. O'Brien and Fleming suggested

$$\alpha_{OF}(\tau) = 2\left[1 - \Phi\left(\frac{z_\alpha}{\sqrt{\tau}}\right)\right].$$

Equivalence studies

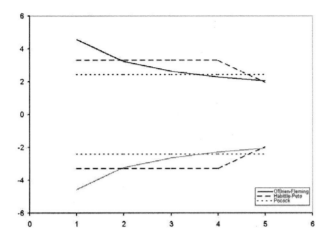

Figure 10.3: Different approaches to interim analysis in a clinical trial: z values for significance.

The method by Pocock can also be used. There,

$$\alpha_P(\tau) = \alpha \log\left[1 + \tau(e-1)\right].$$

Alternatively, alpha can be distributed evenly (uniformly) over the information fraction:

$$\alpha_U(\tau) = \alpha \cdot \tau.$$

Figure 10.4 displays the value of α as a function of the information fraction, for the different methods described here.

10.5 Equivalence studies

Statisticians have learned, during their training, that testing of a statistical hypothesis can lead to two different conclusions. If we wish to test the hypothesis H_o: $\mu_1 = \mu_2$ against H_1: $\mu_1 \neq \mu_2$, one of the following two results may emerge:

1. The test statistic is "significant". This means that we have obtained a result which, if H_o is true, carries a low probability. This would lead us to reject H_o in favour of H_1.

2. The test statistic is "not significant". This means that we do not have enough evidence to reject H_o.

Table 10.5: Some proposed stopping rules for interim analyses.

Interim analysis number	O'Brien-Fleming		Haybittle-Pareto		Pocock	
	z	p	z	p	z	p
			$R = 2$			
1	2.782	0.0054	2.576	0.0100	2.178	0.0294
2	1.967	0.0492	1.960	0.0500	2.178	0.0294
			$R = 3$			
1	3.438	0.0006	2.576	0.0100	2.289	0.0221
2	2.431	0.0151	2.576	0.0100	2.289	0.0221
3	1.985	0.0471	1.960	0.0500	2.289	0.0221
			$R = 4$			
1	4.084	$5 \cdot 10^{-5}$	3.291	0.0010	2.361	0.0182
2	2.888	0.0039	3.291	0.0010	2.361	0.0182
3	2.358	0.0184	3.291	0.0010	2.361	0.0182
4	2.042	0.0412	1.960	0.0500	2.361	0.0182
			$R = 5$			
1	4.555	$5 \cdot 10^{-6}$	3.291	0.0010	2.413	0.0158
2	3.221	0.0013	3.291	0.0010	2.413	0.0158
3	2.630	0.0085	3.291	0.0010	2.413	0.0158
4	2.277	0.0228	3.291	0.0010	2.413	0.0158
5	2.037	0.0417	1.960	0.0500	2.413	0.0158

There is an important philosophical distinction between these two cases. In the first case, we make a firm decision: we decide that H_o can be rejected. Although there is still some doubt (after all, the value of the test statistic is only improbable, not impossible, if H_o is true), we are prepared to take that risk. A significant test statistic leads to a decision to reject H_o in favour of H_1.

In the second case, where H_o cannot be rejected, we are not in a position to take any strong decisions. The failure to reject H_o may be caused by a number of reasons: the "true" difference may be very small; the testing procedure may be too weak to detect the difference; the sample sizes may be too small, and so on. Thus, frequentist hypothesis testing can never lead to accepting H_o. All we can say is that H_o cannot be rejected.

Equivalence studies

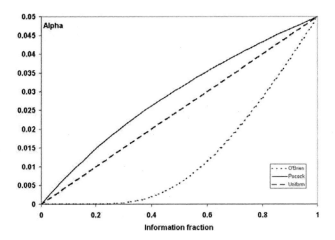

Figure 10.4: Alpha spending functions according to O'Brien-Fleming, Pocock, and the uniform approach.

10.5.1 "Clinically important differences"

However, in clinical trials, there is often interest in demonstrating that a new drug is "equivalent" to an existing drug. If the drugs are equivalent the clinical effect, as measured by some variable y, should be "the same" for both drugs. This means that we need to design a clinical study aiming at "proving" the null hypothesis.

One way of demonstrating equivalence is to set up a so called Clinically important difference that we wish to detect. Denote the means of the two variables y_1 and y_2 with μ_1 and μ_2, respectively. Denote the difference between the mean values with $\tau = \mu_1 - \mu_2$. For the moment we assume that the standard deviations are the same, equal to σ, for the two variables. Choose a value δ of a clinically important difference such that a difference in mean value between the treatments that is larger than δ indicates that the treatments cannot be considered equivalent, while a difference smaller than δ is interpreted as "equivalence".

We now wish to show that $-\delta < \tau < \delta$. This can be done by performing two single-sided hypothesis tests:

$$H_{0A} : \quad \tau \geq \delta \text{ against } H_{1A}: \tau < \delta$$
$$H_{0B} : \quad \tau \leq -\delta \text{ against } H_{1B}: \tau > -\delta$$

If we cannot reject H_{0A}, the two drugs are not equivalent. If we cannot

10 Topics in clinical trials

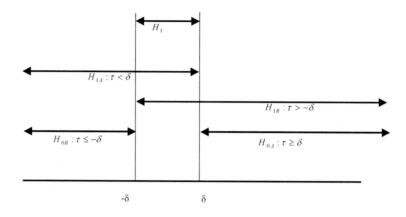

Figure 10.5: Rejection regions for different hypotheses in an equivalence study; figure based on Piantadosi (1997).

reject H_{0B} they are not equivalent either. But if both the null hypotheses can be rejected, there is some evidence that the difference between the two treatments is "small": using our criterion, they may be considered to be equivalent.

Thus, the hypothesis of equivalence corresponds to the following:

$$H_0 : H_{0A} \cup H_{0B}$$
$$H_1 : H_{1A} \cap H_{1B}$$

If H_0 can be rejected, we have been able to demonstrate equivalence, using this definition. If each of the tests is made at level $\alpha/2$, the joint test H_0 is made at a level close to α. This is equivalent to stating that the entire confidence interval for the difference τ should be within the indifference region $\pm\delta$; see Figure 10.5.

Example 10.10 *We want to compare a new drug with an existing standard drug. One hundred patients are randomized to each of the drugs, and the effect is measured. Results:*

Drug 1   Drug 2
$\bar{y}_1 = 100$ $\bar{y}_2 = 97$
$s_1 = 6$ $s_2 = 6$
$n_1 = 100$ $n_2 = 100$

Can the two drugs be considered equivalent if we define equivalence such that the difference in mean value between the drugs should not exceed

$\delta = 5$?

We first test H_{0A}: $\mu_1 - \mu_2 \geq 5$ against H_{1A}: $\mu_1 - \mu_2 < 5$ using a two-sample t test. The pooled estimate of variance is $\hat{\sigma}^2 = 36$. We use the 2.5% level, i.e. we should reject H_0 if $t < -1.97$, which is the 2.5% limit for t on 198 df. The test statistic is

$$t = \frac{100 - 97 - 5}{\sqrt{36\left(\frac{1}{100} + \frac{1}{100}\right)}} = \frac{-2}{\sqrt{\frac{36 \cdot 2}{100}}} = -2.357.$$

Thus, we can reject H_{0A}.

We go on to test H_{0B}: $\mu_1 - \mu_2 \leq -5$ against H_{1B}: $\mu_1 - \mu_2 > -5$. The rejection region is $t > 1.97$. The test statistic is

$$t = \frac{100 - 97 + 5}{\sqrt{36\left(\frac{1}{100} + \frac{1}{100}\right)}} = \frac{8}{\sqrt{\frac{36 \cdot 2}{100}}} = 9.428.$$

This is clearly significant and we reject H_{0B}. Since we have rejected both H_{0A} and H_{0B}, we conclude that the two drugs are equivalent, using the definition that their mean values should differ by less than $\delta = 5$ units.

10.5.2 Superiority and inferiority

In some cases, one may wish to demonstrate "equivalence" in one direction only. This means that we want to conclude that a new treatment is "not worse" than the old treatment, where we define "not worse" as "at most δ units worse". Note that the hypotheses tested are then

$$\begin{aligned} H_0 &: \tau \geq \delta \\ H_1 &: \tau < \delta \end{aligned}$$

This is in contrast with the set of hypotheses used to demonstrate that the new treatment is better than the old one. If H_0 can be rejected, we have demonstrated non-inferiority.

10.5.3 Equivalence studies with proportions

The same set of arguments used above for mean values apply also on other parameters, such as proportions. In fact, if we want to compare two proportions, we want to test hypotheses about the parameter $p_1 - p_2$. If we denote $\tau = p_1 - p_2$, we can use exactly the same statements about hypotheses as we did above; the only difference is that the test statistics used to test the hypotheses are different. Large-sample tests comparing two proportions can be made using normal approximation. Alternatively,

χ^2 tests, with or without continuity correction, can be used. If the samples are small, Fisher's exact test can be used. Any of these tests can be used to study equivalence for proportions.

10.5.4 Sample size determination

Sample size for mean values

Determination of the sample size needed to demonstrate equivalence is very sensitive to the choice of δ. Based on similar arguments as for standard tests, the sample size needed to demonstrate equivalence in a two-sample t test situation is

$$n = \frac{2\sigma^2 (z_{1-\alpha} + z_{1-\beta})^2}{[\delta - (\mu_1 - \mu_2)]^2}.$$

It is common to set $\mu_1 = \mu_2$ in this expression.

Example 10.11 *Suppose that we want to compare two treatments to show that the new treatment is "equivalent" to the old one. We assume that the old treatment will have a mean value around 100 and a standard deviation of 10 units. We are willing to regard the new treatment as equivalent if its mean value is within ±5 units from that of the old treatment. The test will be made at the 5% level and we want a power of 90%. The sample size needed to achieve this is*

$$n = \frac{2 \cdot 10^2 (1.96 + 1.282)^2}{5^2} = 84.1.$$

We need at least 85 patients in each group.

Sample size for proportions

In a similar way, the sample size needed to demonstrate equivalence for proportions is

$$n = \frac{[p_1 (1 - p_1) + p_2 (1 - p_2)] (z_{1-\alpha} + z_{1-\beta})^2}{[\delta - (p_1 - p_2)]^2}.$$

Also here, it is often assumed that $p_1 = p_2$.

Example 10.12 *Suppose that the traditional treatment has a cure rate of 50% for some disease. We are willing to regard a new treatment as equivalent if it has a cure rate of between 40% and 60%, i.e. if $\delta = 0.1$.*

We will use a test with $\alpha = 0.05$ and want the test to have a power of 90%. The sample size needed to achieve this is

$$n = \frac{[0.5(1-0.5) + 0.5(1-0.5)](1.96 + 1.282)^2}{[0.10]^2} = 525.5.$$

We need at least 526 patients in each group. If we change δ to 0.15, the required sample size is reduced to 234 in each group, so the sample size is sensitive to the choice of δ.

10.6 Dose-finding studies

The object of dose-finding studies is to find the dose of a drug that results in a certain reaction among a certain fraction of the patients. The reaction might be recovery, improvement, adverse reactions, etc. Thus, some types of reactions may be positive to the patient while some are negative. In many cases (such as cytostatica), the dose-finding study should find a compromise between the adverse effects and the therapeutic effect. Dose-finding studies are Phase I-II studies, i.e. they are made at a rather early stage in the clinical testing of a new drug.

From a methodological point of view, dose-finding studies are related to dose-response studies that are often made on animals. We start the discussion by briefly describing dose-response studies. Then, the modifications necessary, for ethical reasons, when the experiment is made on humans is discussed.

10.6.1 Dose-response studies

In a typical dose-response study, batches of animals are exposed to a given set of doses of some compound. After some time, the number of animals in each batch that have experienced a certain reaction (death, recovery, etc.) are counted. As an example, Finney (1947) studied the relation between Rotenone concentration and % affected animals. The data are given in the following table:

Conc	Log(Conc)	No. of insects	No. affected	% affected
10.2	1.01	50	44	88
7.7	0.89	49	42	86
5.1	0.71	46	24	52
3.8	0.58	48	16	33
2.6	0.41	50	6	12

10 Topics in clinical trials

The relation between $x = \text{Log(Conc)}$ and $\widehat{p} = \%$ affected is often modeled as a Logistic function or as a cumulative Normal function. The first case is called *logit analysis* while the second is called *probit analysis*. For the Logit model,

$$\text{logit}(\widehat{p}) = \log \frac{\widehat{p}}{1-\widehat{p}} = \beta_0 + \beta_1 x,$$

and for the probit model

$$\Phi^{-1}(\widehat{p}) = \beta_0 + \beta_1 x.$$

Estimation of the parameters of models of this type can be made using generalized linear models. Once the parameters have been estimated, the estimates can be used for prediction. For both models, an estimate of the log(dose) leading to a 50% reaction is (see e.g. Olsson, 2002, p 92)

$$D_{50} = -\frac{\widehat{\beta}_0}{\widehat{\beta}_1}.$$

The optimum design, for a logit or probit model is to choose two log(dose) values symmetrically placed around the D_{50} value. Such designs, with just a few doses, do not allow model checking. Instead, in practice, a set of doses is often chosen such that the log(dose) values are equidistant and such that the smallest dose gives a reaction just above 0 and the largest dose a reaction close to 1.

10.6.2 Dose-finding studies

In studies on humans, it is not ethically defendable to use a design that includes a full range of doses; this might cause unacceptable side affects among patients who receive high doses. Instead, dose-finding studies are adaptive. This means that the trials start with a safely low dose, and that the choice of dose for the next patient or group of patients depends on the result from the previous ones. It is common to try each dose on a group of three patients before the next dose is tried.

The goal, of course, is to find a dose that will not harm the patient, while at the same time it has therapeutic benefits. As part of this investigation, one seeks to establish the Maximum Tolerable Dose (MTD). The level of toxicity or side effects that is acceptable is called the dose-limiting toxicity (DLT). The DLT of a drug typically depends on the disease the drug is intended to cure. Thus, for example, cancer treatments can be acceptable although they produce a great deal of side effects. Thus, assessing the DLT is a clinical, rather than a statistical, task.

The MTD is often determined using a binary outcome (presence or absence of DLT) (Storer, 1998). The dose X_{MTD} at which DLT occurs is the dose for which

$$P(DLT|X_{MTD}) = q_0$$

The level of q_0 is often not defined. In practice, however, q_0 is (implicitly) defined to be about 0.25.

Dose escalation

One way to choose the starting dose is to use one tenth of the dose that causes a 10% mortality in rats (Piantadosi, 1997). If no side effects are encountered on a group of patients using this dose, the next higher dose is used. An interesting way to decide the doses to use is to use a Fibonacci series:

1, 1, 2, 3, 5, 8, 13, 21,...

where each term is the sum of the two preceding terms; the ratio between succeeding terms converges towards the "golden ratio", about 1.62.. In practice, the doses are often selected to increase at a slightly slower rate than the Fibonacci series. Sequences such as

1, 2, 1.67, 1.5, 1.4, 1.3, 1.3, 1.3, ...

are often used instead. Sometimes, escalation proceeds logarithmically.

The dose escalation follows an algorithm that has been decided before the trial. An example of a dose escalation algorithm is presented in Figure 10.6. The dose level below the one where dose escalation was stopped is used as X_{MTD}.

The procedure described in Figure 10.6 is rather complex, from the point of view of statistical inference. The procedure is sequential, and the number of iterations is not known beforehand. Since sample sizes are often small, the statistical power may be rather low. A few other approaches have been described in the literature and we will briefly summarize a few of them.

10.6.3 A Bayesian approach: Continual reassessment

Since the amount of data is scarce in this type of studies, use of Bayesian methods is a tempting alternative. O'Quigley et al (1991) suggested a Bayesian approach that is based on the following ideas.

10 Topics in clinical trials

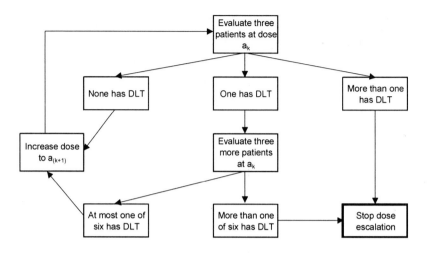

Figure 10.6: An example of a dose escalation algorithm. The graph is based on the description in Storer (1998).

Using a series of dose levels such as $[a_0, a_1, \ldots]$, start at the dose level that is believed to give a probability of DLT close to q_0. A simple dose-response model ψ is used. In fact, it is suggested that the model should be a one-parameter model like

$$\psi(x, \theta) = [(\tanh x + 1)/2]^\theta. \tag{10.1}$$

This excludes dose-response models like the logistic one. In models of type (10.1) it is assumed that there exists a unique θ_0 such that

$$\psi(x_{MTD}, \theta_0) = q_0.$$

A prior distribution $g(\theta)$ is assumed for the parameter θ. It should be chosen such that, for the chosen starting dose a,

$$\int_0^\infty \psi(a, \theta) g(\theta) \, d\theta = q_0,$$

or

$$\psi(a, \mu_a) = q_0$$

where μ_a is the mean in the prior. After each patient, the distribution $g(\theta)$ is updated. Each new patient is treated with the dose that minimizes some measure of distance between the current estimate of toxicity and the target q_0. When the last patient has been studied, the next version of $g(\theta)$ is used to point estimate x_{MTD}. Formulas for confidence intervals for x_{MTD} are still not available.

The method sketched above is open to the same debate as other Bayesian methods. The results depend on the choice of prior.

10.6.4 Storer's two-stage design

Storer (1989) has suggested a two-stage procedure for determination of x_{MTD}. It is based on the view that as much sampling effort as possible should be used at doses close to x_{MTD}. It works as follows:

1. In the first stage, single patients are used at each dose level. The dose level is increased until the first toxicity is reached.

2. In the second stage, a dose level one step below the last level in step 1 is used. A fixed number of groups of three patients are entered. For each group: If none experiences DLT, the dose is stepped up. If one experiences DLT, the dose is retained. If two or more experience DLT, the dose is reduced.

When all groups have been entered, a dose-response model (such as the logistic model $\log \frac{p}{1-p} = \beta_0 + \beta_1 x$) is fitted to the data. The estimated MTD is then

$$x_{MTD} = \text{logit}\left(q_0 - \widehat{\beta}_0\right)/\widehat{\beta}_1.$$

Storer notes that the described procedure works best when $q_0 \approx 0.33$. The fact that the procedure is sequential means that standard Maximum Likelihood confidence intervals tend to be anti-conservative. More correct confidence intervals can be obtained by simulation, but these tend to be rather wide.

10 Topics in clinical trials

10.7 Exercises

Exercise 10.1 *The cholesterol level (mg/dl) was measured twice for each of ten patients. A one-way Anova was run on the data Results:*

```
Analysis of Variance for y, using Adjusted SS for Tests
Source     DF    Seq SS    Adj SS    Adj MS      F        P
Patient     9   1231.46   1231.46    136.83   110.64   0.000
Error      10     12.37     12.37      1.24
Total      19   1243.82
```

A. Calculate the reliability of the cholesterol measurements.

B. A study is being planned to compare the effects of two diets on the cholesterol level. The study should have a 90% probability of detecting a mean difference of 10 mg/dl, using a t test (double sided test at the 5% level). Only one measurement will be made on each patient. How large sample is needed?

C. Suppose that you only measure whether the diet increases the cholesterol level or not, i.e. a binary response. How large sample is needed to obtain a 90% power for a double sided test of the hypothesis $H_0: p_1 = p_2$ if the true difference is $p_1 - p_2 = 0.1$?

Exercise 10.2 *The following data come from a reliability study where the same variable was measured five times on each of four persons.*

Person	n	$\bar{y}$	s^2
1	5	10.4	5.2
2	5	14.5	6.7
3	5	12.2	5.7
4	5	11.8	6.1

A. Estimate the reliability coefficient

B. Two treatments will be compared in an experiment. It is desired that the test should have a power of 0.9 for a test at the 5% level, if the true mean difference is 1 unit. How large sample is needed?

Exercise 10.3 *A research team works with a clinical trial concerned with treatments against an eye disease. The trial includes two treatments, A and B, where B is the new treatment. The trial is planned such that a total of $n = 120$ patients will be recruited. The patients are randomized on the two treatments such that the two treatment groups are balanced in size at each time point.*

Interim analyses will be undertaken at two occasions during the trial: after 40 patients have been treated and after 80 patients have been treated.

The primary endpoint is whether the patient has been cured ($y = 1$) or not ($y = 0$). All analyses will be made as double sided tests, using some method that accounts for the fact that interim analyses are made.

Question:

A. After 40 patients have been treated, it turns out that $x = 11$ patients in group A and $y = 15$ patients n group B are cured. Should the trial continue? Conclusion?

B. After 80 patients have been treated, it turns out that $x = 24$ patients in group A and $y = 32$ patients in group B have been cured. Should the trial continue? Conclusion?

C. When the whole trial is completed ($n = 120$) it turns out that $x = 35$ patients in group A and $y = 46$ patients in group B are cured. Conclusion?

D. Briefly comment the choice of method for the tests.

Exercise 10.4 In a clinical trial with two treatments, interim analyses will be made at three time points: after 50 patients; after 100 patients and at the end of the trial after 150 patients. The following summary data were obtained are the three occasions. For each time point, state whether the experiment should continue and which conclusion can be drawn. Double sided tests at a joint 5% level should be used.

A. Time point 1:

$$\bar{y}_1 = 104 \quad s_1 = 10 \quad n_1 = 25 \quad \bar{y}_2 = 100 \quad s_2 = 8 \quad n_2 = 25$$

B. Time point 2

$$\bar{y}_1 = 103 \quad s_1 = 8 \quad n_1 = 50 \quad \bar{y}_2 = 99 \quad s_2 = 9 \quad n_2 = 50$$

C. Time point 3

$$\bar{y}_1 = 102 \quad s_1 = 10 \quad n_1 = 75 \quad \bar{y}_2 = 98 \quad s_2 = 11 \quad n_2 = 75$$

D. Briefly comment on the choice of method for the interim analyses.

Exercise 10.5 In some types of clinical trials, fir example analysis of survival, it is difficult to plan the trial such that interim analyses can be made at sample sizes decided beforehand. In such cases, it may be better to make interim analyses based on the "alpha spending function" approach. In survival analysis, you can calculate an "information fraction", which is the cumulative number of deaths observed at the time of the analysis, divided by the expected total number of deaths during the study.

10 Topics in clinical trials

In an experiment[1], the beta blocker Propraniol has been compared with placebo for patients who recently suffered a cardiac infarction. The patients would be followed for three years, with follow-up every half year. The information fraction for each time point will be the cumulative number of new infarctions, divided by the total number of patients ($n = 3837$). Pocock's method for "alpha spending function" will be used. After each follow-up, some test statistic Z that can be regarded as standard Normal, is calculated, Results:

Period	Cases	Test statistic Z
1	131	1.68
2	59	2.24
3	688	2.37
4	796	2.30
5	853	2.34
6	904	2.82

Question: For each period, calculate the p value the test will be made at, using Pocock's method for alpha spending function. For each period, decide whether the experiment should be stopped, based on this p value and the reported Z values. The total level of the test should be $\alpha=0.05$ (double sided test).

Exercise 10.6 After a cardiac infarction, there is a risk that the patient may get new infarctions. Therefore, different types of treatments are used immediately after the first attack. Two such treatments will be compared in a clinical study: Sulfinpyrazone (S) and a new treatment (N). Sulfinpyrazone has certain side effects that are absent in the new drug. It is therefore of interest to examine whether the newt treatment N may be regarded as equivalent with S. The relapse percentage with S is usually around 10% within a year after the first infarction. A difference between the drugs of $\pm 3\%$ means that the drugs are similar enough to be considered equivalent with regard to relapse percentage.

A randomized clinical study has been conducted. All patients were followed during one year after the first infarction and it was recorded whether the patient had a new infarction or not. Results:

Treatment	New infarction	No infarction	Total
S	61	631	692
N	57	625	682
Total	118	1256	1374

---

[1] DeMets et al: Statistical aspects of early termination of the Betablocker heart attack trial. Controlled clinical trials, 5:362-372, 1984.

Questions
A. Test, on the level $\alpha=0.10$, whether the two treatments may be regarded as equivalent.
B. Assume that a new study is being planned, to examine the equivalence of S and N. Equivalence is defined as a difference of $\pm 3\%$ and the relapse percentage is around 10%. The test of equivalence will be made at the level $\alpha=0.10$. How large sample is needed for the test to have a power of at least $1-\beta=0.9$?

Exercise 10.7 A randomized clinical study has been made to examine whether two treatments against a chronic disease may be regarded as equivalent. The primary endpoint is whether or not the patient improved during the study. The treatments may be regarded as equivalent if the difference in improvement rate is within $\pm 3\%$. Results:

Result	A	B	Total
Not improved	38	24	62
Improved	172	162	334
	210	186	396

Questions:
A. Test at the level $\alpha=0.05$ whether the treatments may be regarded as equivalent.
B. Assume that a new study will be planned to examine the equivalence of A and B. We define equivalence as a difference of at most $\pm 3\%$. Further, we assume that an estimate of the proportion improved may be taken from question A. How large sample is needed if the equivalence test should have a power of $1-\beta=0.9$. The test will be made at level $\alpha=0.05$.

Exercise 10.8 Consider the dose escalation procedure described in Figure 10.6. The probability that a patient has DLT (dose-limiting toxicity) with the dose used is 0.25 and this is the value where the increase in dose should be stopped.
A. What is the probability that the dose escalation is stopped?
B. Give a formula for the probability that dose escalation is stopped, given that the probability of DLT is p.
C. Make of graph of the probability of DLT as a function of p, for $p = 0.0$, $p = 0.25$, $p = 0.5$, $p = 0.75$ and $p = 1.0$.

Tables

Table 1: Random numbers

Rad	Kolumn 1 2 3 4 5	6 7 8 9 10	11 12 13 14 15	16 17 18 19 20	21 22 23 24 25	26 27 28 29 30
1	9 8 5 5 1	6 3 5 0 9	0 1 2 8 7	5 9 8 9 9	3 3 6 6 1	7 0 4 6 2
2	3 9 5 5 0	4 6 9 2 2	8 1 5 8 7	0 9 3 2 9	8 7 9 3 8	3 6 1 4 0
3	1 1 5 1 2	0 6 7 2 7	5 3 7 1 1	8 5 2 8 4	6 1 8 0 8	6 3 8 0 9
4	5 9 2 3 2	9 2 8 1 7	3 2 5 3 0	8 5 8 2 9	0 2 0 0 0	5 3 3 3 1
5	1 4 0 8 3	3 9 5 2 5	3 9 5 8 0	5 8 7 0 3	9 6 6 6 5	7 2 2 2 3
6	7 6 7 7 0	9 6 8 7 7	8 6 0 3 7	6 8 6 0 9	9 3 6 7 2	1 6 0 1 8
7	5 2 7 5 5	9 6 4 8 9	9 1 1 0 4	8 4 0 6 8	6 9 7 0 2	6 7 4 7 3
8	0 5 8 7 4	0 3 3 0 1	6 5 4 0 9	0 7 3 7 9	3 3 5 4 7	4 4 0 5 2
9	2 1 2 6 2	5 8 7 0 9	5 6 8 8 7	6 7 0 7 2	3 5 0 8 3	2 5 4 0 9
10	5 3 3 0 8	6 8 8 4 1	7 7 1 0 2	0 7 6 2 2	4 1 4 1 6	9 3 8 4 4
11	6 9 8 6 9	3 2 8 5 7	3 6 9 9 9	2 4 7 6 3	9 9 5 8 9	1 4 7 5 4
12	5 8 9 6 0	6 5 2 4 3	7 9 9 3 2	5 0 2 7 2	5 1 8 9 3	7 0 3 3 2
13	8 0 7 5 6	6 2 4 8 0	3 4 0 2 8	6 0 7 7 2	9 5 6 2 6	8 5 3 8 4
14	6 5 6 1 7	2 0 2 5 4	8 3 1 1 7	0 7 5 2 9	8 4 5 7 2	9 3 3 5 2
15	6 0 3 5 5	7 0 6 6 9	3 3 6 8 0	7 8 5 6 1	3 3 0 4 4	1 4 8 0 2
16	6 9 7 8 2	3 8 4 6 8	0 8 5 3 1	0 4 7 6 2	3 4 4 3 6	4 3 0 2 5
17	5 7 5 3 3	2 9 1 4 0	6 3 1 5 2	0 6 6 9 3	6 6 6 6 8	9 8 1 1 0
18	0 8 5 4 5	2 9 2 2 8	2 4 8 5 9	8 3 7 6 3	5 0 1 0 7	3 2 9 9 6
19	0 3 4 1 4	8 4 7 1 5	5 7 8 5 1	0 4 3 7 7	6 5 9 2 6	1 5 9 2 9
20	2 0 8 2 2	6 4 5 3 4	0 4 6 1 2	0 8 3 7 0	3 1 1 8 7	8 4 6 2 3
21	6 4 9 5 1	4 1 6 6 4	5 9 3 4 7	6 3 4 2 4	1 3 2 0 1	4 3 4 3 7
22	2 3 8 3 9	6 8 7 1 4	7 2 1 9 5	0 8 3 1 5	2 4 1 1 6	3 9 6 2 0
23	5 7 6 0 4	9 4 8 2 7	4 4 1 9 2	3 0 2 7 4	2 2 8 7 9	9 3 9 3 6
24	3 8 6 5 0	6 0 3 2 3	5 1 5 3 5	8 4 8 9 4	3 2 1 6 1	1 9 1 1 6
25	8 7 9 9 3	1 0 4 5 7	8 3 0 9 6	9 8 9 4 9	8 3 5 4 7	6 4 3 3 6
26	5 7 8 3 1	9 3 3 8 2	9 1 0 4 8	4 8 9 5 8	0 0 4 0 1	6 8 5 1 8
27	1 5 2 4 9	3 5 1 7 3	5 0 1 6 4	2 5 3 0 3	4 3 9 3 5	3 1 6 4 0
28	7 1 5 2 7	4 5 4 7 8	5 2 0 7 2	9 8 9 2 9	5 4 5 9 4	2 7 0 9 4
29	3 8 9 8 5	2 9 3 1 1	1 4 3 8 3	3 5 7 0 1	1 2 2 4 1	7 8 6 6 1
30	8 9 5 9 3	8 9 0 2 1	0 0 8 6 5	1 3 8 5 4	0 3 1 0 4	4 2 1 0 1
31	6 7 2 6 3	8 3 8 0 3	4 8 1 5 5	2 6 3 0 9	3 9 9 2 5	2 2 1 0 5
32	1 1 8 6 6	7 5 2 8 4	8 4 3 3 4	6 3 9 8 0	0 9 9 7 8	5 9 4 1 8
33	1 7 6 3 1	9 0 5 4 4	1 4 8 8 6	1 9 1 1 7	1 8 6 7 8	8 7 1 1 6
34	2 1 1 8 7	7 3 4 6 1	3 4 4 9 6	3 6 8 7 7	9 8 8 4 4	7 0 8 2 0
35	5 4 9 4 8	7 1 8 8 8	9 8 2 1 8	0 1 6 2 1	8 2 4 5 4	9 1 5 5 9
36	8 3 3 7 2	9 6 5 1 6	7 4 6 6 5	7 3 6 2 9	1 1 3 5 6	2 0 0 7 9
37	5 8 6 1 7	4 7 9 3 4	8 6 2 7 2	9 3 8 5 6	7 1 9 5 6	3 7 8 5 9
38	4 5 9 4 4	6 7 2 0 3	8 6 8 7 4	3 9 7 4 7	5 4 5 0 2	1 3 9 7 5
39	3 2 4 6 9	4 9 5 9 6	9 4 2 2 3	6 1 2 6 7	0 5 1 8 1	4 9 1 0 0
40	4 0 1 8 2	6 1 0 8 3	3 7 1 7 5	7 4 9 4 9	1 3 7 5 2	4 4 7 2 4
41	5 8 2 6 2	8 5 2 2 1	1 0 7 0 4	1 9 2 3 1	2 5 3 0 1	5 5 4 0 1
42	2 0 4 3 1	8 1 9 4 0	6 9 6 3 7	1 4 9 8 7	2 4 8 8 0	2 1 0 8 9
43	4 7 4 8 4	6 7 6 4 5	4 9 0 0 2	6 9 2 1 3	2 8 2 1 2	2 8 4 5 3
44	2 1 6 7 8	5 6 2 8 2	5 0 9 6 2	2 8 4 1 0	1 7 9 9 1	7 2 0 4 4
45	6 9 9 8 0	9 5 4 3 1	7 4 8 7 9	1 1 8 3 2	1 1 9 3 1	0 7 3 0 1

Table 2: Distribution function of the Binomial distribution

n	x	p=0.1	p=0.2	p=0.25	p=0.3	p=0.4	p=0.5	p=0.6	p=0.7	p=0.75	p=0.8	p=0.9
5	0	0.5905	0.3277	0.2373	0.1681	0.0778	0.0313	0.0102	0.0024	0.0010	0.0003	0.0000
	1	0.9185	0.7373	0.6328	0.5282	0.3370	0.1875	0.0870	0.0308	0.0156	0.0067	0.0005
	2	0.9914	0.9421	0.8965	0.8369	0.6826	0.5000	0.3174	0.1631	0.1035	0.0579	0.0086
	3	0.9995	0.9933	0.9844	0.9692	0.9130	0.8125	0.6630	0.4718	0.3672	0.2627	0.0815
	4	1.0000	0.9997	0.9990	0.9976	0.9898	0.9688	0.9222	0.8319	0.7627	0.6723	0.4095
	5	1.0000	1.0000	1.0000	1.0000	1.0000	1.0000	1.0000	1.0000	1.0000	1.0000	1.0000
6	0	0.5314	0.2621	0.1780	0.1176	0.0467	0.0156	0.0041	0.0007	0.0002	0.0001	0.0000
	1	0.8857	0.6554	0.5339	0.4202	0.2333	0.1094	0.0410	0.0109	0.0046	0.0016	0.0001
	2	0.9842	0.9011	0.8306	0.7443	0.5443	0.3438	0.1792	0.0705	0.0376	0.0170	0.0013
	3	0.9987	0.9830	0.9624	0.9295	0.8208	0.6562	0.4557	0.2557	0.1694	0.0989	0.0159
	4	0.9999	0.9984	0.9954	0.9891	0.9590	0.8906	0.7667	0.5798	0.4661	0.3446	0.1143
	5	1.0000	0.9999	0.9998	0.9993	0.9959	0.9844	0.9533	0.8824	0.8220	0.7379	0.4686
	6	1.0000	1.0000	1.0000	1.0000	1.0000	1.0000	1.0000	1.0000	1.0000	1.0000	1.0000
7	0	0.4783	0.2097	0.1335	0.0824	0.0280	0.0078	0.0016	0.0002	0.0001	0.0000	0.0000
	1	0.8503	0.5767	0.4449	0.3294	0.1586	0.0625	0.0188	0.0038	0.0013	0.0004	0.0000
	2	0.9743	0.8520	0.7564	0.6471	0.4199	0.2266	0.0963	0.0288	0.0129	0.0047	0.0002
	3	0.9973	0.9667	0.9294	0.8740	0.7102	0.5000	0.2898	0.1260	0.0706	0.0333	0.0027
	4	0.9998	0.9953	0.9871	0.9712	0.9037	0.7734	0.5801	0.3529	0.2436	0.1480	0.0257
	5	1.0000	0.9996	0.9987	0.9962	0.9812	0.9375	0.8414	0.6706	0.5551	0.4233	0.1497
	6	1.0000	1.0000	0.9999	0.9998	0.9984	0.9922	0.9720	0.9176	0.8665	0.7903	0.5217
	7	1.0000	1.0000	1.0000	1.0000	1.0000	1.0000	1.0000	1.0000	1.0000	1.0000	1.0000
8	0	0.4305	0.1678	0.1001	0.0576	0.0168	0.0039	0.0007	0.0001	0.0000	0.0000	0.0000
	1	0.8131	0.5033	0.3671	0.2553	0.1064	0.0352	0.0085	0.0013	0.0004	0.0001	0.0000
	2	0.9619	0.7969	0.6785	0.5518	0.3154	0.1445	0.0498	0.0113	0.0042	0.0012	0.0000
	3	0.9950	0.9437	0.8862	0.8059	0.5941	0.3633	0.1737	0.0580	0.0273	0.0104	0.0004
	4	0.9996	0.9896	0.9727	0.9420	0.8263	0.6367	0.4059	0.1941	0.1138	0.0563	0.0050
	5	1.0000	0.9988	0.9958	0.9887	0.9502	0.8555	0.6846	0.4482	0.3215	0.2031	0.0381
	6	1.0000	0.9999	0.9996	0.9987	0.9915	0.9648	0.8936	0.7447	0.6329	0.4967	0.1869
	7	1.0000	1.0000	1.0000	0.9999	0.9993	0.9961	0.9832	0.9424	0.8999	0.8322	0.5695
	8	1.0000	1.0000	1.0000	1.0000	1.0000	1.0000	1.0000	1.0000	1.0000	1.0000	1.0000
9	0	0.3874	0.1342	0.0751	0.0404	0.0101	0.0020	0.0003	0.0000	0.0000	0.0000	0.0000
	1	0.7748	0.4362	0.3003	0.1960	0.0705	0.0195	0.0038	0.0004	0.0001	0.0000	0.0000
	2	0.9470	0.7382	0.6007	0.4628	0.2318	0.0898	0.0250	0.0043	0.0013	0.0003	0.0000
	3	0.9917	0.9144	0.8343	0.7297	0.4826	0.2539	0.0994	0.0253	0.0100	0.0031	0.0001
	4	0.9991	0.9804	0.9511	0.9012	0.7334	0.5000	0.2666	0.0988	0.0489	0.0196	0.0009
	5	0.9999	0.9969	0.9900	0.9747	0.9006	0.7461	0.5174	0.2703	0.1657	0.0856	0.0083
	6	1.0000	0.9997	0.9987	0.9957	0.9750	0.9102	0.7682	0.5372	0.3993	0.2618	0.0530
	7	1.0000	1.0000	0.9999	0.9996	0.9962	0.9805	0.9295	0.8040	0.6997	0.5638	0.2252
	8	1.0000	1.0000	1.0000	1.0000	0.9997	0.9980	0.9899	0.9596	0.9249	0.8658	0.6126
	9	1.0000	1.0000	1.0000	1.0000	1.0000	1.0000	1.0000	1.0000	1.0000	1.0000	1.0000
10	0	0.3487	0.1074	0.0563	0.0282	0.0060	0.0010	0.0001	0.0000	0.0000	0.0000	0.0000
	1	0.7361	0.3758	0.2440	0.1493	0.0464	0.0107	0.0017	0.0001	0.0000	0.0000	0.0000
	2	0.9298	0.6778	0.5256	0.3828	0.1673	0.0547	0.0123	0.0016	0.0004	0.0001	0.0000
	3	0.9872	0.8791	0.7759	0.6496	0.3823	0.1719	0.0548	0.0106	0.0035	0.0009	0.0000
	4	0.9984	0.9672	0.9219	0.8497	0.6331	0.3770	0.1662	0.0473	0.0197	0.0064	0.0001
	5	0.9999	0.9936	0.9803	0.9527	0.8338	0.6230	0.3669	0.1503	0.0781	0.0328	0.0016
	6	1.0000	0.9991	0.9965	0.9894	0.9452	0.8281	0.6177	0.3504	0.2241	0.1209	0.0128
	7	1.0000	0.9999	0.9996	0.9984	0.9877	0.9453	0.8327	0.6172	0.4744	0.3222	0.0702
	8	1.0000	1.0000	1.0000	0.9999	0.9983	0.9893	0.9536	0.8507	0.7560	0.6242	0.2639
	9	1.0000	1.0000	1.0000	1.0000	0.9999	0.9990	0.9940	0.9718	0.9437	0.8926	0.6513
	10	1.0000	1.0000	1.0000	1.0000	1.0000	1.0000	1.0000	1.0000	1.0000	1.0000	1.0000

n	x	p=0.1	p=0.2	p=0.25	p=0.3	p=0.4	p=0.5	p=0.6	p=0.7	p=0.75	p=0.8	p=0.9
11	0	0.3138	0.0859	0.0422	0.0198	0.0036	0.0005	0.0000	0.0000	0.0000	0.0000	0.0000
	1	0.6974	0.3221	0.1971	0.1130	0.0302	0.0059	0.0007	0.0000	0.0000	0.0000	0.0000
	2	0.9104	0.6174	0.4552	0.3127	0.1189	0.0327	0.0059	0.0006	0.0001	0.0000	0.0000
	3	0.9815	0.8389	0.7133	0.5696	0.2963	0.1133	0.0293	0.0043	0.0012	0.0002	0.0000
	4	0.9972	0.9496	0.8854	0.7897	0.5328	0.2744	0.0994	0.0216	0.0076	0.0020	0.0000
	5	0.9997	0.9883	0.9657	0.9218	0.7535	0.5000	0.2465	0.0782	0.0343	0.0117	0.0003
	6	1.0000	0.9980	0.9924	0.9784	0.9006	0.7256	0.4672	0.2103	0.1146	0.0504	0.0028
	7	1.0000	0.9998	0.9988	0.9957	0.9707	0.8867	0.7037	0.4304	0.2867	0.1611	0.0185
	8	1.0000	1.0000	0.9999	0.9994	0.9941	0.9673	0.8811	0.6873	0.5448	0.3826	0.0896
	9	1.0000	1.0000	1.0000	1.0000	0.9993	0.9941	0.9698	0.8870	0.8029	0.6779	0.3026
	10	1.0000	1.0000	1.0000	1.0000	1.0000	0.9995	0.9964	0.9802	0.9578	0.9141	0.6862
	11	1.0000	1.0000	1.0000	1.0000	1.0000	1.0000	1.0000	1.0000	1.0000	1.0000	1.0000
12	0	0.2824	0.0687	0.0317	0.0138	0.0022	0.0002	0.0000	0.0000	0.0000	0.0000	0.0000
	1	0.6590	0.2749	0.1584	0.0850	0.0196	0.0032	0.0003	0.0000	0.0000	0.0000	0.0000
	2	0.8891	0.5583	0.3907	0.2528	0.0834	0.0193	0.0028	0.0002	0.0000	0.0000	0.0000
	3	0.9744	0.7946	0.6488	0.4925	0.2253	0.0730	0.0153	0.0017	0.0004	0.0001	0.0000
	4	0.9957	0.9274	0.8424	0.7237	0.4382	0.1938	0.0573	0.0095	0.0028	0.0006	0.0000
	5	0.9995	0.9806	0.9456	0.8822	0.6652	0.3872	0.1582	0.0386	0.0143	0.0039	0.0001
	6	0.9999	0.9961	0.9857	0.9614	0.8418	0.6128	0.3348	0.1178	0.0544	0.0194	0.0005
	7	1.0000	0.9994	0.9972	0.9905	0.9427	0.8062	0.5618	0.2763	0.1576	0.0726	0.0043
	8	1.0000	0.9999	0.9996	0.9983	0.9847	0.9270	0.7747	0.5075	0.3512	0.2054	0.0256
	9	1.0000	1.0000	1.0000	0.9998	0.9972	0.9807	0.9166	0.7472	0.6093	0.4417	0.1109
	10	1.0000	1.0000	1.0000	1.0000	0.9997	0.9968	0.9804	0.9150	0.8416	0.7251	0.3410
	11	1.0000	1.0000	1.0000	1.0000	1.0000	0.9998	0.9978	0.9862	0.9683	0.9313	0.7176
	12	1.0000	1.0000	1.0000	1.0000	1.0000	1.0000	1.0000	1.0000	1.0000	1.0000	1.0000
13	0	0.2542	0.0550	0.0238	0.0097	0.0013	0.0001	0.0000	0.0000	0.0000	0.0000	0.0000
	1	0.6213	0.2336	0.1267	0.0637	0.0126	0.0017	0.0001	0.0000	0.0000	0.0000	0.0000
	2	0.8661	0.5017	0.3326	0.2025	0.0579	0.0112	0.0013	0.0001	0.0000	0.0000	0.0000
	3	0.9658	0.7473	0.5843	0.4206	0.1686	0.0461	0.0078	0.0007	0.0001	0.0000	0.0000
	4	0.9935	0.9009	0.7940	0.6543	0.3530	0.1334	0.0321	0.0040	0.0010	0.0002	0.0000
	5	0.9991	0.9700	0.9198	0.8346	0.5744	0.2905	0.0977	0.0182	0.0056	0.0012	0.0000
	6	0.9999	0.9930	0.9757	0.9376	0.7712	0.5000	0.2288	0.0624	0.0243	0.0070	0.0001
	7	1.0000	0.9988	0.9944	0.9818	0.9023	0.7095	0.4256	0.1654	0.0802	0.0300	0.0009
	8	1.0000	0.9998	0.9990	0.9960	0.9679	0.8666	0.6470	0.3457	0.2060	0.0991	0.0065
	9	1.0000	1.0000	0.9999	0.9993	0.9922	0.9539	0.8314	0.5794	0.4157	0.2527	0.0342
	10	1.0000	1.0000	1.0000	0.9999	0.9987	0.9888	0.9421	0.7975	0.6674	0.4983	0.1339
	11	1.0000	1.0000	1.0000	1.0000	0.9999	0.9983	0.9874	0.9363	0.8733	0.7664	0.3787
	12	1.0000	1.0000	1.0000	1.0000	1.0000	0.9999	0.9987	0.9903	0.9762	0.9450	0.7458
	13	1.0000	1.0000	1.0000	1.0000	1.0000	1.0000	1.0000	1.0000	1.0000	1.0000	1.0000
14	0	0.2288	0.0440	0.0178	0.0068	0.0008	0.0001	0.0000	0.0000	0.0000	0.0000	0.0000
	1	0.5846	0.1979	0.1010	0.0475	0.0081	0.0009	0.0001	0.0000	0.0000	0.0000	0.0000
	2	0.8416	0.4481	0.2811	0.1608	0.0398	0.0065	0.0006	0.0000	0.0000	0.0000	0.0000
	3	0.9559	0.6982	0.5213	0.3552	0.1243	0.0287	0.0039	0.0002	0.0000	0.0000	0.0000
	4	0.9908	0.8702	0.7415	0.5842	0.2793	0.0898	0.0175	0.0017	0.0003	0.0000	0.0000
	5	0.9985	0.9561	0.8883	0.7805	0.4859	0.2120	0.0583	0.0083	0.0022	0.0004	0.0000
	6	0.9998	0.9884	0.9617	0.9067	0.6925	0.3953	0.1501	0.0315	0.0103	0.0024	0.0000
	7	1.0000	0.9976	0.9897	0.9685	0.8499	0.6047	0.3075	0.0933	0.0383	0.0116	0.0002
	8	1.0000	0.9996	0.9978	0.9917	0.9417	0.7880	0.5141	0.2195	0.1117	0.0439	0.0015
	9	1.0000	1.0000	0.9997	0.9983	0.9825	0.9102	0.7207	0.4158	0.2585	0.1298	0.0092
	10	1.0000	1.0000	1.0000	0.9998	0.9961	0.9713	0.8757	0.6448	0.4787	0.3018	0.0441
	11	1.0000	1.0000	1.0000	1.0000	0.9994	0.9935	0.9602	0.8392	0.7189	0.5519	0.1584
	12	1.0000	1.0000	1.0000	1.0000	0.9999	0.9991	0.9919	0.9525	0.8990	0.8021	0.4154
	13	1.0000	1.0000	1.0000	1.0000	1.0000	0.9999	0.9992	0.9932	0.9822	0.9560	0.7712
	14	1.0000	1.0000	1.0000	1.0000	1.0000	1.0000	1.0000	1.0000	1.0000	1.0000	1.0000

n	x	p=0.1	p=0.2	p=0.25	p=0.3	p=0.4	p=0.5	p=0.6	p=0.7	p=0.75	p=0.8	p=0.9
15	0	0.2059	0.0352	0.0134	0.0047	0.0005	0.0000	0.0000	0.0000	0.0000	0.0000	0.0000
	1	0.5490	0.1671	0.0802	0.0353	0.0052	0.0005	0.0000	0.0000	0.0000	0.0000	0.0000
	2	0.8159	0.3980	0.2361	0.1268	0.0271	0.0037	0.0003	0.0000	0.0000	0.0000	0.0000
	3	0.9444	0.6482	0.4613	0.2969	0.0905	0.0176	0.0019	0.0001	0.0000	0.0000	0.0000
	4	0.9873	0.8358	0.6865	0.5155	0.2173	0.0592	0.0093	0.0007	0.0001	0.0000	0.0000
	5	0.9978	0.9389	0.8516	0.7216	0.4032	0.1509	0.0338	0.0037	0.0008	0.0001	0.0000
	6	0.9997	0.9819	0.9434	0.8689	0.6098	0.3036	0.0950	0.0152	0.0042	0.0008	0.0000
	7	1.0000	0.9958	0.9827	0.9500	0.7869	0.5000	0.2131	0.0500	0.0173	0.0042	0.0000
	8	1.0000	0.9992	0.9958	0.9848	0.9050	0.6964	0.3902	0.1311	0.0566	0.0181	0.0003
	9	1.0000	0.9999	0.9992	0.9963	0.9662	0.8491	0.5968	0.2784	0.1484	0.0611	0.0022
	10	1.0000	1.0000	0.9999	0.9993	0.9907	0.9408	0.7827	0.4845	0.3135	0.1642	0.0127
	11	1.0000	1.0000	1.0000	0.9999	0.9981	0.9824	0.9095	0.7031	0.5387	0.3518	0.0556
	12	1.0000	1.0000	1.0000	1.0000	0.9997	0.9963	0.9729	0.8732	0.7639	0.6020	0.1841
	13	1.0000	1.0000	1.0000	1.0000	1.0000	0.9995	0.9948	0.9647	0.9198	0.8329	0.4510
	14	1.0000	1.0000	1.0000	1.0000	1.0000	1.0000	0.9995	0.9953	0.9866	0.9648	0.7941
	15	1.0000	1.0000	1.0000	1.0000	1.0000	1.0000	1.0000	1.0000	1.0000	1.0000	1.0000
16	0	0.1853	0.0281	0.0100	0.0033	0.0003	0.0000	0.0000	0.0000	0.0000	0.0000	0.0000
	1	0.5147	0.1407	0.0635	0.0261	0.0033	0.0003	0.0000	0.0000	0.0000	0.0000	0.0000
	2	0.7892	0.3518	0.1971	0.0994	0.0183	0.0021	0.0001	0.0000	0.0000	0.0000	0.0000
	3	0.9316	0.5981	0.4050	0.2459	0.0651	0.0106	0.0009	0.0000	0.0000	0.0000	0.0000
	4	0.9830	0.7982	0.6302	0.4499	0.1666	0.0384	0.0049	0.0003	0.0000	0.0000	0.0000
	5	0.9967	0.9183	0.8103	0.6598	0.3288	0.1051	0.0191	0.0016	0.0003	0.0000	0.0000
	6	0.9995	0.9733	0.9204	0.8247	0.5272	0.2272	0.0583	0.0071	0.0016	0.0002	0.0000
	7	0.9999	0.9930	0.9729	0.9256	0.7161	0.4018	0.1423	0.0257	0.0075	0.0015	0.0000
	8	1.0000	0.9985	0.9925	0.9743	0.8577	0.5982	0.2839	0.0744	0.0271	0.0070	0.0001
	9	1.0000	0.9998	0.9984	0.9929	0.9417	0.7728	0.4728	0.1753	0.0796	0.0267	0.0005
	10	1.0000	1.0000	0.9997	0.9984	0.9809	0.8949	0.6712	0.3402	0.1897	0.0817	0.0033
	11	1.0000	1.0000	1.0000	0.9997	0.9951	0.9616	0.8334	0.5501	0.3698	0.2018	0.0170
	12	1.0000	1.0000	1.0000	1.0000	0.9991	0.9894	0.9349	0.7541	0.5950	0.4019	0.0684
	13	1.0000	1.0000	1.0000	1.0000	0.9999	0.9979	0.9817	0.9006	0.8029	0.6482	0.2108
	14	1.0000	1.0000	1.0000	1.0000	1.0000	0.9997	0.9967	0.9739	0.9365	0.8593	0.4853
	15	1.0000	1.0000	1.0000	1.0000	1.0000	1.0000	0.9997	0.9967	0.9900	0.9719	0.8147
	16	1.0000	1.0000	1.0000	1.0000	1.0000	1.0000	1.0000	1.0000	1.0000	1.0000	1.0000
17	0	0.1668	0.0225	0.0075	0.0023	0.0002	0.0000	0.0000	0.0000	0.0000	0.0000	0.0000
	1	0.4818	0.1182	0.0501	0.0193	0.0021	0.0001	0.0000	0.0000	0.0000	0.0000	0.0000
	2	0.7618	0.3096	0.1637	0.0774	0.0123	0.0012	0.0001	0.0000	0.0000	0.0000	0.0000
	3	0.9174	0.5489	0.3530	0.2019	0.0464	0.0064	0.0005	0.0000	0.0000	0.0000	0.0000
	4	0.9779	0.7582	0.5739	0.3887	0.1260	0.0245	0.0025	0.0001	0.0000	0.0000	0.0000
	5	0.9953	0.8943	0.7653	0.5968	0.2639	0.0717	0.0106	0.0007	0.0001	0.0000	0.0000
	6	0.9992	0.9623	0.8929	0.7752	0.4478	0.1662	0.0348	0.0032	0.0006	0.0001	0.0000
	7	0.9999	0.9891	0.9598	0.8954	0.6405	0.3145	0.0919	0.0127	0.0031	0.0005	0.0000
	8	1.0000	0.9974	0.9876	0.9597	0.8011	0.5000	0.1989	0.0403	0.0124	0.0026	0.0000
	9	1.0000	0.9995	0.9969	0.9873	0.9081	0.6855	0.3595	0.1046	0.0402	0.0109	0.0001
	10	1.0000	0.9999	0.9994	0.9968	0.9652	0.8338	0.5522	0.2248	0.1071	0.0377	0.0008
	11	1.0000	1.0000	0.9999	0.9993	0.9894	0.9283	0.7361	0.4032	0.2347	0.1057	0.0047
	12	1.0000	1.0000	1.0000	0.9999	0.9975	0.9755	0.8740	0.6113	0.4261	0.2418	0.0221
	13	1.0000	1.0000	1.0000	1.0000	0.9995	0.9936	0.9536	0.7981	0.6470	0.4511	0.0826
	14	1.0000	1.0000	1.0000	1.0000	0.9999	0.9988	0.9877	0.9226	0.8363	0.6904	0.2382
	15	1.0000	1.0000	1.0000	1.0000	1.0000	0.9999	0.9979	0.9807	0.9499	0.8818	0.5182
	16	1.0000	1.0000	1.0000	1.0000	1.0000	1.0000	0.9998	0.9977	0.9925	0.9775	0.8332
	17	1.0000	1.0000	1.0000	1.0000	1.0000	1.0000	1.0000	1.0000	1.0000	1.0000	1.0000

n	x	p=0.1	p=0.2	p=0.25	p=0.3	p=0.4	p=0.5	p=0.6	p=0.7	p=0.75	p=0.8	p=0.9
18	0	0.1501	0.0180	0.0056	0.0016	0.0001	0.0000	0.0000	0.0000	0.0000	0.0000	0.0000
	1	0.4503	0.0991	0.0395	0.0142	0.0013	0.0001	0.0000	0.0000	0.0000	0.0000	0.0000
	2	0.7338	0.2713	0.1353	0.0600	0.0082	0.0007	0.0000	0.0000	0.0000	0.0000	0.0000
	3	0.9018	0.5010	0.3057	0.1646	0.0328	0.0038	0.0002	0.0000	0.0000	0.0000	0.0000
	4	0.9718	0.7164	0.5187	0.3327	0.0942	0.0154	0.0013	0.0000	0.0000	0.0000	0.0000
	5	0.9936	0.8671	0.7175	0.5344	0.2088	0.0481	0.0058	0.0003	0.0000	0.0000	0.0000
	6	0.9988	0.9487	0.8610	0.7217	0.3743	0.1189	0.0203	0.0014	0.0002	0.0000	0.0000
	7	0.9998	0.9837	0.9431	0.8593	0.5634	0.2403	0.0576	0.0061	0.0012	0.0002	0.0000
	8	1.0000	0.9957	0.9807	0.9404	0.7368	0.4073	0.1347	0.0210	0.0054	0.0009	0.0000
	9	1.0000	0.9991	0.9946	0.9790	0.8653	0.5927	0.2632	0.0596	0.0193	0.0043	0.0000
	10	1.0000	0.9998	0.9988	0.9939	0.9424	0.7597	0.4366	0.1407	0.0569	0.0163	0.0002
	11	1.0000	1.0000	0.9998	0.9986	0.9797	0.8811	0.6257	0.2783	0.1390	0.0513	0.0012
	12	1.0000	1.0000	1.0000	0.9997	0.9942	0.9519	0.7912	0.4656	0.2825	0.1329	0.0064
	13	1.0000	1.0000	1.0000	1.0000	0.9987	0.9846	0.9058	0.6673	0.4813	0.2836	0.0282
	14	1.0000	1.0000	1.0000	1.0000	0.9998	0.9962	0.9672	0.8354	0.6943	0.4990	0.0982
	15	1.0000	1.0000	1.0000	1.0000	1.0000	0.9993	0.9918	0.9400	0.8647	0.7287	0.2662
	16	1.0000	1.0000	1.0000	1.0000	1.0000	0.9999	0.9987	0.9858	0.9605	0.9009	0.5497
	17	1.0000	1.0000	1.0000	1.0000	1.0000	1.0000	0.9999	0.9984	0.9944	0.9820	0.8499
	18	1.0000	1.0000	1.0000	1.0000	1.0000	1.0000	1.0000	1.0000	1.0000	1.0000	1.0000
19	0	0.1351	0.0144	0.0042	0.0011	0.0001	0.0000	0.0000	0.0000	0.0000	0.0000	0.0000
	1	0.4203	0.0829	0.0310	0.0104	0.0008	0.0000	0.0000	0.0000	0.0000	0.0000	0.0000
	2	0.7054	0.2369	0.1113	0.0462	0.0055	0.0004	0.0000	0.0000	0.0000	0.0000	0.0000
	3	0.8850	0.4551	0.2631	0.1332	0.0230	0.0022	0.0001	0.0000	0.0000	0.0000	0.0000
	4	0.9648	0.6733	0.4654	0.2822	0.0696	0.0096	0.0006	0.0000	0.0000	0.0000	0.0000
	5	0.9914	0.8369	0.6678	0.4739	0.1629	0.0318	0.0031	0.0001	0.0000	0.0000	0.0000
	6	0.9983	0.9324	0.8251	0.6655	0.3081	0.0835	0.0116	0.0006	0.0001	0.0000	0.0000
	7	0.9997	0.9767	0.9225	0.8180	0.4878	0.1796	0.0352	0.0028	0.0005	0.0000	0.0000
	8	1.0000	0.9933	0.9713	0.9161	0.6675	0.3238	0.0885	0.0105	0.0023	0.0003	0.0000
	9	1.0000	0.9984	0.9911	0.9674	0.8139	0.5000	0.1861	0.0326	0.0089	0.0016	0.0000
	10	1.0000	0.9997	0.9977	0.9895	0.9115	0.6762	0.3325	0.0839	0.0287	0.0067	0.0000
	11	1.0000	1.0000	0.9995	0.9972	0.9648	0.8204	0.5122	0.1820	0.0775	0.0233	0.0003
	12	1.0000	1.0000	0.9999	0.9994	0.9884	0.9165	0.6919	0.3345	0.1749	0.0676	0.0017
	13	1.0000	1.0000	1.0000	0.9999	0.9969	0.9682	0.8371	0.5261	0.3322	0.1631	0.0086
	14	1.0000	1.0000	1.0000	1.0000	0.9994	0.9904	0.9304	0.7178	0.5346	0.3267	0.0352
	15	1.0000	1.0000	1.0000	1.0000	0.9999	0.9978	0.9770	0.8668	0.7369	0.5449	0.1150
	16	1.0000	1.0000	1.0000	1.0000	1.0000	0.9996	0.9945	0.9538	0.8887	0.7631	0.2946
	17	1.0000	1.0000	1.0000	1.0000	1.0000	1.0000	0.9992	0.9896	0.9690	0.9171	0.5797
	18	1.0000	1.0000	1.0000	1.0000	1.0000	1.0000	0.9999	0.9989	0.9958	0.9856	0.8649
	19	1.0000	1.0000	1.0000	1.0000	1.0000	1.0000	1.0000	1.0000	1.0000	1.0000	1.0000
20	0	0.1216	0.0115	0.0032	0.0008	0.0000	0.0000	0.0000	0.0000	0.0000	0.0000	0.0000
	1	0.3917	0.0692	0.0243	0.0076	0.0005	0.0000	0.0000	0.0000	0.0000	0.0000	0.0000
	2	0.6769	0.2061	0.0913	0.0355	0.0036	0.0002	0.0000	0.0000	0.0000	0.0000	0.0000
	3	0.8670	0.4114	0.2252	0.1071	0.0160	0.0013	0.0000	0.0000	0.0000	0.0000	0.0000
	4	0.9568	0.6296	0.4148	0.2375	0.0510	0.0059	0.0003	0.0000	0.0000	0.0000	0.0000
	5	0.9887	0.8042	0.6172	0.4164	0.1256	0.0207	0.0016	0.0000	0.0000	0.0000	0.0000
	6	0.9976	0.9133	0.7858	0.6080	0.2500	0.0577	0.0065	0.0003	0.0000	0.0000	0.0000
	7	0.9996	0.9679	0.8982	0.7723	0.4159	0.1316	0.0210	0.0013	0.0002	0.0000	0.0000
	8	0.9999	0.9900	0.9591	0.8867	0.5956	0.2517	0.0565	0.0051	0.0009	0.0001	0.0000
	9	1.0000	0.9974	0.9861	0.9520	0.7553	0.4119	0.1275	0.0171	0.0039	0.0006	0.0000
	10	1.0000	0.9994	0.9961	0.9829	0.8725	0.5881	0.2447	0.0480	0.0139	0.0026	0.0000
	11	1.0000	0.9999	0.9991	0.9949	0.9435	0.7483	0.4044	0.1133	0.0409	0.0100	0.0001
	12	1.0000	1.0000	0.9998	0.9987	0.9790	0.8684	0.5841	0.2277	0.1018	0.0321	0.0004
	13	1.0000	1.0000	1.0000	0.9997	0.9935	0.9423	0.7500	0.3920	0.2142	0.0867	0.0024
	14	1.0000	1.0000	1.0000	1.0000	0.9984	0.9793	0.8744	0.5836	0.3828	0.1958	0.0113
	15	1.0000	1.0000	1.0000	1.0000	0.9997	0.9941	0.9490	0.7625	0.5852	0.3704	0.0432
	16	1.0000	1.0000	1.0000	1.0000	1.0000	0.9987	0.9840	0.8929	0.7748	0.5886	0.1330
	17	1.0000	1.0000	1.0000	1.0000	1.0000	0.9998	0.9964	0.9645	0.9087	0.7939	0.3231
	18	1.0000	1.0000	1.0000	1.0000	1.0000	1.0000	0.9995	0.9924	0.9757	0.9308	0.6083
	19	1.0000	1.0000	1.0000	1.0000	1.0000	1.0000	1.0000	0.9992	0.9968	0.9885	0.8784
	20	1.0000	1.0000	1.0000	1.0000	1.0000	1.0000	1.0000	1.0000	1.0000	1.0000	1.0000

Table 3: Distribution function of the Poisson distribution

x	λ=0.5	λ=1	λ=2	λ=3	λ=4	λ=5	λ=6	λ=7	λ=8	λ=9
0	0.607	0.368	0.135	0.050	0.018	0.007	0.002	0.001	0.000	0.000
1	0.910	0.736	0.406	0.199	0.092	0.040	0.017	0.007	0.003	0.001
2	0.986	0.920	0.677	0.423	0.238	0.125	0.062	0.030	0.014	0.006
3	0.998	0.981	0.857	0.647	0.433	0.265	0.151	0.082	0.042	0.021
4	1.000	0.996	0.947	0.815	0.629	0.440	0.285	0.173	0.100	0.055
5		0.999	0.983	0.916	0.785	0.616	0.446	0.301	0.191	0.116
6		1.000	0.995	0.966	0.889	0.762	0.606	0.450	0.313	0.207
7			0.999	0.988	0.949	0.867	0.744	0.599	0.453	0.324
8			1.000	0.996	0.979	0.932	0.847	0.729	0.593	0.456
9				0.999	0.992	0.968	0.916	0.830	0.717	0.587
10				1.000	0.997	0.986	0.957	0.901	0.816	0.706
11					0.999	0.995	0.980	0.947	0.888	0.803
12					1.000	0.998	0.991	0.973	0.936	0.876
13						0.999	0.996	0.987	0.966	0.926
14						1.000	0.999	0.994	0.983	0.959
15							0.999	0.998	0.992	0.978
16							1.000	0.999	0.996	0.989
17								1.000	0.998	0.995
18									0.999	0.998
19									1.000	0.999
20										1.000

x	λ=10	λ=11	λ=12	λ=13	λ=14	λ=15
0	0.000	0.000	0.000	0.000	0.000	0.000
1	0.000	0.000	0.000	0.000	0.000	0.000
2	0.003	0.001	0.001	0.000	0.000	0.000
3	0.010	0.005	0.002	0.001	0.000	0.000
4	0.029	0.015	0.008	0.004	0.002	0.001
5	0.067	0.038	0.020	0.011	0.006	0.003
6	0.130	0.079	0.046	0.026	0.014	0.008
7	0.220	0.143	0.090	0.054	0.032	0.018
8	0.333	0.232	0.155	0.100	0.062	0.037
9	0.458	0.341	0.242	0.166	0.109	0.070
10	0.583	0.460	0.347	0.252	0.176	0.118
11	0.697	0.579	0.462	0.353	0.260	0.185
12	0.792	0.689	0.576	0.463	0.358	0.268
13	0.864	0.781	0.682	0.573	0.464	0.363
14	0.917	0.854	0.772	0.675	0.570	0.466
15	0.951	0.907	0.844	0.764	0.669	0.568
16	0.973	0.944	0.899	0.835	0.756	0.664
17	0.986	0.968	0.937	0.890	0.827	0.749
18	0.993	0.982	0.963	0.930	0.883	0.819
19	0.997	0.991	0.979	0.957	0.923	0.875
20	0.998	0.995	0.988	0.975	0.952	0.917
21	0.999	0.998	0.994	0.986	0.971	0.947
22	1.000	0.999	0.997	0.992	0.983	0.967
23		1.000	0.999	0.996	0.991	0.981
24			0.999	0.998	0.995	0.989
25			1.000	0.999	0.997	0.994
26				1.000	0.999	0.997
27					0.999	0.998
28					1.000	0.999
29						1.000
30						

Table 4: Distribution function of the Normal distribution

z	0	1	2	3	4	5	6	7	8	9
0.0	0.5000	0.5040	0.5080	0.5120	0.5160	0.5199	0.5239	0.5279	0.5319	0.5359
0.1	0.5398	0.5438	0.5478	0.5517	0.5557	0.5596	0.5636	0.5675	0.5714	0.5753
0.2	0.5793	0.5832	0.5871	0.5910	0.5948	0.5987	0.6026	0.6064	0.6103	0.6141
0.3	0.6179	0.6217	0.6255	0.6293	0.6331	0.6368	0.6406	0.6443	0.6480	0.6517
0.4	0.6554	0.6591	0.6628	0.6664	0.6700	0.6736	0.6772	0.6808	0.6844	0.6879
0.5	0.6915	0.6950	0.6985	0.7019	0.7054	0.7088	0.7123	0.7157	0.7190	0.7224
0.6	0.7257	0.7291	0.7324	0.7357	0.7389	0.7422	0.7454	0.7486	0.7517	0.7549
0.7	0.7580	0.7611	0.7642	0.7673	0.7704	0.7734	0.7764	0.7794	0.7823	0.7852
0.8	0.7881	0.7910	0.7939	0.7967	0.7995	0.8023	0.8051	0.8078	0.8106	0.8133
0.9	0.8159	0.8186	0.8212	0.8238	0.8264	0.8289	0.8315	0.8340	0.8365	0.8389
1.0	0.8413	0.8438	0.8461	0.8485	0.8508	0.8531	0.8554	0.8577	0.8599	0.8621
1.1	0.8643	0.8665	0.8686	0.8708	0.8729	0.8749	0.8770	0.8790	0.8810	0.8830
1.2	0.8849	0.8869	0.8888	0.8907	0.8925	0.8944	0.8962	0.8980	0.8997	0.9015
1.3	0.9032	0.9049	0.9066	0.9082	0.9099	0.9115	0.9131	0.9147	0.9162	0.9177
1.4	0.9192	0.9207	0.9222	0.9236	0.9251	0.9265	0.9279	0.9292	0.9306	0.9319
1.5	0.9332	0.9345	0.9357	0.9370	0.9382	0.9394	0.9406	0.9418	0.9429	0.9441
1.6	0.9452	0.9463	0.9474	0.9484	0.9495	0.9505	0.9515	0.9525	0.9535	0.9545
1.7	0.9554	0.9564	0.9573	0.9582	0.9591	0.9599	0.9608	0.9616	0.9625	0.9633
1.8	0.9641	0.9649	0.9656	0.9664	0.9671	0.9678	0.9686	0.9693	0.9699	0.9706
1.9	0.9713	0.9719	0.9726	0.9732	0.9738	0.9744	0.9750	0.9756	0.9761	0.9767
2.0	0.9772	0.9778	0.9783	0.9788	0.9793	0.9798	0.9803	0.9808	0.9812	0.9817
2.1	0.9821	0.9826	0.9830	0.9834	0.9838	0.9842	0.9846	0.9850	0.9854	0.9857
2.2	0.9861	0.9864	0.9868	0.9871	0.9875	0.9878	0.9881	0.9884	0.9887	0.9890
2.3	0.9893	0.9896	0.9898	0.9901	0.9904	0.9906	0.9909	0.9911	0.9913	0.9916
2.4	0.9918	0.9920	0.9922	0.9925	0.9927	0.9929	0.9931	0.9932	0.9934	0.9936
2.5	0.9938	0.9940	0.9941	0.9943	0.9945	0.9946	0.9948	0.9949	0.9951	0.9952
2.6	0.9953	0.9955	0.9956	0.9957	0.9959	0.9960	0.9961	0.9962	0.9963	0.9964
2.7	0.9965	0.9966	0.9967	0.9968	0.9969	0.9970	0.9971	0.9972	0.9973	0.9974
2.8	0.9974	0.9975	0.9976	0.9977	0.9977	0.9978	0.9979	0.9979	0.9980	0.9981
2.9	0.9981	0.9982	0.9982	0.9983	0.9984	0.9984	0.9985	0.9985	0.9986	0.9986
3.0	0.9987	0.9987	0.9987	0.9988	0.9988	0.9989	0.9989	0.9989	0.9990	0.9990
3.1	0.9990	0.9991	0.9991	0.9991	0.9992	0.9992	0.9992	0.9992	0.9993	0.9993
3.2	0.9993	0.9993	0.9994	0.9994	0.9994	0.9994	0.9994	0.9995	0.9995	0.9995
3.3	0.9995	0.9995	0.9995	0.9996	0.9996	0.9996	0.9996	0.9996	0.9996	0.9997
3.4	0.9997	0.9997	0.9997	0.9997	0.9997	0.9997	0.9997	0.9997	0.9997	0.9998
3.5	0.9998	0.9998	0.9998	0.9998	0.9998	0.9998	0.9998	0.9998	0.9998	0.9998
3.6	0.9998	0.9998	0.9999	0.9999	0.9999	0.9999	0.9999	0.9999	0.9999	0.9999
3.7	0.9999	0.9999	0.9999	0.9999	0.9999	0.9999	0.9999	0.9999	0.9999	0.9999
3.8	0.9999	0.9999	0.9999	0.9999	0.9999	0.9999	0.9999	0.9999	0.9999	0.9999
3.9	1.0000	1.0000	1.0000	1.0000	1.0000	1.0000	1.0000	1.0000	1.0000	1.0000
4.0	1.0000	1.0000	1.0000	1.0000	1.0000	1.0000	1.0000	1.0000	1.0000	1.0000

The table presents the standard Normal distribution function $F(z) = \int_{-\infty}^{z} \frac{1}{\sqrt{2\pi}} e^{-t^2/2} dt$ for values of z in the interval $z=0.00(0.01)4.09$. For negative values of z use the relation $F(z) = 1 - F(-z)$.

Table 5: t distribution function

d.f.	0.75	0.90	0.95	0.975	F(t) 0.99	0.995	0.9975	0.999	0.9995
1	1.000	3.078	6.314	12.706	31.821	63.657	127.321	318.309	636.619
2	0.816	1.886	2.920	4.303	6.965	9.925	14.089	22.327	31.599
3	0.765	1.638	2.353	3.182	4.541	5.841	7.453	10.215	12.924
4	0.741	1.533	2.132	2.776	3.747	4.604	5.598	7.173	8.610
5	0.727	1.476	2.015	2.571	3.365	4.032	4.773	5.893	6.869
6	0.718	1.440	1.943	2.447	3.143	3.707	4.317	5.208	5.959
7	0.711	1.415	1.895	2.365	2.998	3.499	4.029	4.785	5.408
8	0.706	1.397	1.860	2.306	2.896	3.355	3.833	4.501	5.041
9	0.703	1.383	1.833	2.262	2.821	3.250	3.690	4.297	4.781
10	0.700	1.372	1.812	2.228	2.764	3.169	3.581	4.144	4.587
11	0.697	1.363	1.796	2.201	2.718	3.106	3.497	4.025	4.437
12	0.695	1.356	1.782	2.179	2.681	3.055	3.428	3.930	4.318
13	0.694	1.350	1.771	2.160	2.650	3.012	3.372	3.852	4.221
14	0.692	1.345	1.761	2.145	2.624	2.977	3.326	3.787	4.140
15	0.691	1.341	1.753	2.131	2.602	2.947	3.286	3.733	4.073
16	0.690	1.337	1.746	2.120	2.583	2.921	3.252	3.686	4.015
17	0.689	1.333	1.740	2.110	2.567	2.898	3.222	3.646	3.965
18	0.688	1.330	1.734	2.101	2.552	2.878	3.197	3.610	3.922
19	0.688	1.328	1.729	2.093	2.539	2.861	3.174	3.579	3.883
20	0.687	1.325	1.725	2.086	2.528	2.845	3.153	3.552	3.850
21	0.686	1.323	1.721	2.080	2.518	2.831	3.135	3.527	3.819
22	0.686	1.321	1.717	2.074	2.508	2.819	3.119	3.505	3.792
23	0.685	1.319	1.714	2.069	2.500	2.807	3.104	3.485	3.768
24	0.685	1.318	1.711	2.064	2.492	2.797	3.091	3.467	3.745
25	0.684	1.316	1.708	2.060	2.485	2.787	3.078	3.450	3.725
26	0.684	1.315	1.706	2.056	2.479	2.779	3.067	3.435	3.707
27	0.684	1.314	1.703	2.052	2.473	2.771	3.057	3.421	3.690
28	0.683	1.313	1.701	2.048	2.467	2.763	3.047	3.408	3.674
29	0.683	1.311	1.699	2.045	2.462	2.756	3.038	3.396	3.659
30	0.683	1.310	1.697	2.042	2.457	2.750	3.030	3.385	3.646
40	0.681	1.303	1.684	2.021	2.423	2.704	2.971	3.307	3.551
50	0.679	1.299	1.676	2.009	2.403	2.678	2.937	3.261	3.496
60	0.679	1.296	1.671	2.000	2.390	2.660	2.915	3.232	3.460
80	0.678	1.292	1.664	1.990	2.374	2.639	2.887	3.195	3.416
100	0.677	1.290	1.660	1.984	2.364	2.626	2.871	3.174	3.390
∞	0.674	1.282	1.645	1.960	2.326	2.576	2.807	3.092	3.291

Example: For a t distribution on 10 degrees of freedom (d.f.) it holds that $P(t \leq 1.812) = 0.95$. The distribution is symmetric so $P(t \leq -1.812) = 0.05$.

Table 6: χ^2 distribution function

d.f.	0.0005	0.001	0.0025	0.005	0.01	0.025	0.05	0.1	0.5
1	0.000	0.000	0.000	0.000	0.000	0.001	0.004	0.016	0.455
2	0.001	0.002	0.005	0.010	0.020	0.051	0.103	0.211	1.386
3	0.015	0.024	0.045	0.072	0.115	0.216	0.352	0.584	2.366
4	0.064	0.091	0.145	0.207	0.297	0.484	0.711	1.064	3.357
5	0.158	0.210	0.307	0.412	0.554	0.831	1.145	1.610	4.351
6	0.299	0.381	0.527	0.676	0.872	1.237	1.635	2.204	5.348
7	0.485	0.598	0.794	0.989	1.239	1.690	2.167	2.833	6.346
8	0.710	0.857	1.104	1.344	1.646	2.180	2.733	3.490	7.344
9	0.972	1.152	1.450	1.735	2.088	2.700	3.325	4.168	8.343
10	1.265	1.479	1.827	2.156	2.558	3.247	3.940	4.865	9.342
11	1.587	1.834	2.232	2.603	3.053	3.816	4.575	5.578	10.341
12	1.934	2.214	2.661	3.074	3.571	4.404	5.226	6.304	11.340
13	2.305	2.617	3.112	3.565	4.107	5.009	5.892	7.042	12.340
14	2.697	3.041	3.582	4.075	4.660	5.629	6.571	7.790	13.339
15	3.108	3.483	4.070	4.601	5.229	6.262	7.261	8.547	14.339
16	3.536	3.942	4.573	5.142	5.812	6.908	7.962	9.312	15.338
17	3.980	4.416	5.092	5.697	6.408	7.564	8.672	10.085	16.338
18	4.439	4.905	5.623	6.265	7.015	8.231	9.390	10.865	17.338
19	4.912	5.407	6.167	6.844	7.633	8.907	10.117	11.651	18.338
20	5.398	5.921	6.723	7.434	8.260	9.591	10.851	12.443	19.337
21	5.896	6.447	7.289	8.034	8.897	10.283	11.591	13.240	20.337
22	6.404	6.983	7.865	8.643	9.542	10.982	12.338	14.041	21.337
23	6.924	7.529	8.450	9.260	10.196	11.689	13.091	14.848	22.337
24	7.453	8.085	9.044	9.886	10.856	12.401	13.848	15.659	23.337
25	7.991	8.649	9.646	10.520	11.524	13.120	14.611	16.473	24.337
26	8.538	9.222	10.256	11.160	12.198	13.844	15.379	17.292	25.336
27	9.093	9.803	10.873	11.808	12.879	14.573	16.151	18.114	26.336
28	9.656	10.391	11.497	12.461	13.565	15.308	16.928	18.939	27.336
29	10.227	10.986	12.128	13.121	14.256	16.047	17.708	19.768	28.336
30	10.804	11.588	12.765	13.787	14.953	16.791	18.493	20.599	29.336
40	16.906	17.916	19.417	20.707	22.164	24.433	26.509	29.051	39.335
50	23.461	24.674	26.464	27.991	29.707	32.357	34.764	37.689	49.335
60	30.340	31.738	33.791	35.534	37.485	40.482	43.188	46.459	59.335
80	44.791	46.520	49.043	51.172	53.540	57.153	60.391	64.278	79.334
100	59.896	61.918	64.857	67.328	70.065	74.222	77.929	82.358	99.334

When the number of degrees of freedom ν is large it holds that $\sqrt{2\chi^2}$ is approximately Normal with mean $\sqrt{2\nu - 1}$ and variance 1. Then, $\chi^2 \approx \frac{1}{2} \left[z + \sqrt{2\nu - 1} \right]^2$.

$F(\chi^2)$

d.f.	0.75	0.9	0.95	0.975	0.99	0.995	0.9975	0.999	0.9995
1	1.323	2.706	3.841	5.024	6.635	7.879	9.141	10.828	12.116
2	2.773	4.605	5.991	7.378	9.210	10.597	11.983	13.816	15.202
3	4.108	6.251	7.815	9.348	11.345	12.838	14.320	16.266	17.730
4	5.385	7.779	9.488	11.143	13.277	14.860	16.424	18.467	19.997
5	6.626	9.236	11.070	12.833	15.086	16.750	18.386	20.515	22.105
6	7.841	10.645	12.592	14.449	16.812	18.548	20.249	22.458	24.103
7	9.037	12.017	14.067	16.013	18.475	20.278	22.040	24.322	26.018
8	10.219	13.362	15.507	17.535	20.090	21.955	23.774	26.124	27.868
9	11.389	14.684	16.919	19.023	21.666	23.589	25.462	27.877	29.666
10	12.549	15.987	18.307	20.483	23.209	25.188	27.112	29.588	31.420
11	13.701	17.275	19.675	21.920	24.725	26.757	28.729	31.264	33.137
12	14.845	18.549	21.026	23.337	26.217	28.300	30.318	32.909	34.821
13	15.984	19.812	22.362	24.736	27.688	29.819	31.883	34.528	36.478
14	17.117	21.064	23.685	26.119	29.141	31.319	33.426	36.123	38.109
15	18.245	22.307	24.996	27.488	30.578	32.801	34.950	37.697	39.719
16	19.369	23.542	26.296	28.845	32.000	34.267	36.456	39.252	41.308
17	20.489	24.769	27.587	30.191	33.409	35.718	37.946	40.790	42.879
18	21.605	25.989	28.869	31.526	34.805	37.156	39.422	42.312	44.434
19	22.718	27.204	30.144	32.852	36.191	38.582	40.885	43.820	45.973
20	23.828	28.412	31.410	34.170	37.566	39.997	42.336	45.315	47.498
21	24.935	29.615	32.671	35.479	38.932	41.401	43.775	46.797	49.011
22	26.039	30.813	33.924	36.781	40.289	42.796	45.204	48.268	50.511
23	27.141	32.007	35.172	38.076	41.638	44.181	46.623	49.728	52.000
24	28.241	33.196	36.415	39.364	42.980	45.559	48.034	51.179	53.479
25	29.339	34.382	37.652	40.646	44.314	46.928	49.435	52.620	54.947
26	30.435	35.563	38.885	41.923	45.642	48.290	50.829	54.052	56.407
27	31.528	36.741	40.113	43.195	46.963	49.645	52.215	55.476	57.858
28	32.620	37.916	41.337	44.461	48.278	50.993	53.594	56.892	59.300
29	33.711	39.087	42.557	45.722	49.588	52.336	54.967	58.301	60.735
30	34.800	40.256	43.773	46.979	50.892	53.672	56.332	59.703	62.162
40	45.616	51.805	55.758	59.342	63.691	66.766	69.699	73.402	76.095
50	56.334	63.167	67.505	71.420	76.154	79.490	82.664	86.661	89.561
60	66.981	74.397	79.082	83.298	88.379	91.952	95.344	99.607	102.695
80	88.130	96.578	101.879	106.629	112.329	116.321	120.102	124.839	128.261
100	109.141	118.498	124.342	129.561	135.807	140.169	144.293	149.449	153.167

Table 7: F distribution, $1 - p = 0.90$

v_2 \ v_1	1	2	3	4	5	6	8	10	12	16	20	40	120
1	39.863	49.500	53.593	55.833	57.240	58.204	59.439	60.195	60.705	61.350	61.740	62.529	63.061
2	8.526	9.000	9.162	9.243	9.293	9.326	9.367	9.392	9.408	9.429	9.441	9.466	9.483
3	5.538	5.462	5.391	5.343	5.309	5.285	5.252	5.230	5.216	5.196	5.184	5.160	5.143
4	4.545	4.325	4.191	4.107	4.051	4.010	3.955	3.920	3.896	3.864	3.844	3.804	3.775
5	4.060	3.780	3.619	3.520	3.453	3.405	3.339	3.297	3.268	3.230	3.207	3.157	3.123
6	3.776	3.463	3.289	3.181	3.108	3.055	2.983	2.937	2.905	2.863	2.836	2.781	2.742
7	3.589	3.257	3.074	2.961	2.883	2.827	2.752	2.703	2.668	2.623	2.595	2.535	2.493
8	3.458	3.113	2.924	2.806	2.726	2.668	2.589	2.538	2.502	2.455	2.425	2.361	2.316
9	3.360	3.006	2.813	2.693	2.611	2.551	2.469	2.416	2.379	2.329	2.298	2.232	2.184
10	3.285	2.924	2.728	2.605	2.522	2.461	2.377	2.323	2.284	2.233	2.201	2.132	2.082
12	3.177	2.807	2.606	2.480	2.394	2.331	2.245	2.188	2.147	2.094	2.060	1.986	1.932
14	3.102	2.726	2.522	2.395	2.307	2.243	2.154	2.095	2.054	1.998	1.962	1.885	1.828
16	3.048	2.668	2.462	2.333	2.244	2.178	2.088	2.028	1.985	1.928	1.891	1.811	1.751
18	3.007	2.624	2.416	2.286	2.196	2.130	2.038	1.977	1.933	1.875	1.837	1.754	1.691
20	2.975	2.589	2.380	2.249	2.158	2.091	1.999	1.937	1.892	1.833	1.794	1.708	1.643
24	2.927	2.538	2.327	2.195	2.103	2.035	1.941	1.877	1.832	1.770	1.730	1.641	1.571
28	2.894	2.503	2.291	2.157	2.064	1.996	1.900	1.836	1.790	1.726	1.685	1.592	1.520
32	2.869	2.477	2.263	2.129	2.036	1.967	1.870	1.805	1.758	1.694	1.652	1.556	1.481
36	2.850	2.456	2.243	2.108	2.014	1.945	1.847	1.781	1.734	1.669	1.626	1.528	1.450
40	2.835	2.440	2.226	2.091	1.997	1.927	1.829	1.763	1.715	1.649	1.605	1.506	1.425
50	2.809	2.412	2.197	2.061	1.966	1.895	1.796	1.729	1.680	1.613	1.568	1.465	1.379
60	2.791	2.393	2.177	2.041	1.946	1.875	1.775	1.707	1.657	1.589	1.543	1.437	1.348
70	2.779	2.380	2.164	2.027	1.931	1.860	1.760	1.691	1.641	1.572	1.526	1.418	1.325
80	2.769	2.370	2.154	2.016	1.921	1.849	1.748	1.680	1.629	1.559	1.513	1.403	1.307
100	2.756	2.356	2.139	2.002	1.906	1.834	1.732	1.663	1.612	1.542	1.494	1.382	1.282

Table 8: F distribution, $1 - p = 0.95$

v_2 \ v_1	1	2	3	4	5	6	8	10	12	16	20	40	120
1	161.45	199.50	215.71	224.58	230.16	233.99	238.88	241.88	243.91	246.46	248.01	251.14	253.25
2	18.513	19.000	19.164	19.247	19.296	19.330	19.371	19.396	19.413	19.433	19.446	19.471	19.487
3	10.128	9.552	9.277	9.117	9.013	8.941	8.845	8.786	8.745	8.692	8.660	8.594	8.549
4	7.709	6.944	6.591	6.388	6.256	6.163	6.041	5.964	5.912	5.844	5.803	5.717	5.658
5	6.608	5.786	5.409	5.192	5.050	4.950	4.818	4.735	4.678	4.604	4.558	4.464	4.398
6	5.987	5.143	4.757	4.534	4.387	4.284	4.147	4.060	4.000	3.922	3.874	3.774	3.705
7	5.591	4.737	4.347	4.120	3.972	3.866	3.726	3.637	3.575	3.494	3.445	3.340	3.267
8	5.318	4.459	4.066	3.838	3.687	3.581	3.438	3.347	3.284	3.202	3.150	3.043	2.967
9	5.117	4.256	3.863	3.633	3.482	3.374	3.230	3.137	3.073	2.989	2.936	2.826	2.748
10	4.965	4.103	3.708	3.478	3.326	3.217	3.072	2.978	2.913	2.828	2.774	2.661	2.580
12	4.747	3.885	3.490	3.259	3.106	2.996	2.849	2.753	2.687	2.599	2.544	2.426	2.341
14	4.600	3.739	3.344	3.112	2.958	2.848	2.699	2.602	2.534	2.445	2.388	2.266	2.178
16	4.494	3.634	3.239	3.007	2.852	2.741	2.591	2.494	2.425	2.333	2.276	2.151	2.059
18	4.414	3.555	3.160	2.928	2.773	2.661	2.510	2.412	2.342	2.250	2.191	2.063	1.968
20	4.351	3.493	3.098	2.866	2.711	2.599	2.447	2.348	2.278	2.184	2.124	1.994	1.896
24	4.260	3.403	3.009	2.776	2.621	2.508	2.355	2.255	2.183	2.088	2.027	1.892	1.790
28	4.196	3.340	2.947	2.714	2.558	2.445	2.291	2.190	2.118	2.021	1.959	1.820	1.714
32	4.149	3.295	2.901	2.668	2.512	2.399	2.244	2.142	2.070	1.972	1.908	1.767	1.657
36	4.113	3.259	2.866	2.634	2.477	2.364	2.209	2.106	2.033	1.934	1.870	1.726	1.612
40	4.085	3.232	2.839	2.606	2.449	2.336	2.180	2.077	2.003	1.904	1.839	1.693	1.577
50	4.034	3.183	2.790	2.557	2.400	2.286	2.130	2.026	1.952	1.850	1.784	1.634	1.511
60	4.001	3.150	2.758	2.525	2.368	2.254	2.097	1.993	1.917	1.815	1.748	1.594	1.467
70	3.978	3.128	2.736	2.503	2.346	2.231	2.074	1.969	1.893	1.790	1.722	1.566	1.435
80	3.960	3.111	2.719	2.486	2.329	2.214	2.056	1.951	1.875	1.772	1.703	1.545	1.411
100	3.936	3.087	2.696	2.463	2.305	2.191	2.032	1.927	1.850	1.746	1.676	1.515	1.376

Table 9: F distribution, $1 - p = 0.99$

v_2 \ v_1	1	2	3	4	5	6	8	10	12	16	20	40	120
1	4052.2	4999.5	5403.4	5624.6	5763.6	5859.0	5981.1	6055.8	6106.3	6170.1	6208.7	6286.8	6339.4
2	98.503	99.000	99.166	99.249	99.299	99.333	99.374	99.399	99.416	99.437	99.449	99.474	99.491
3	34.116	30.817	29.457	28.710	28.237	27.911	27.489	27.229	27.052	26.827	26.690	26.411	26.221
4	21.198	18.000	16.694	15.977	15.522	15.207	14.799	14.546	14.374	14.154	14.020	13.745	13.558
5	16.258	13.274	12.060	11.392	10.967	10.672	10.289	10.051	9.888	9.680	9.553	9.291	9.112
6	13.745	10.925	9.780	9.148	8.746	8.466	8.102	7.874	7.718	7.519	7.396	7.143	6.969
7	12.246	9.547	8.451	7.847	7.460	7.191	6.840	6.620	6.469	6.275	6.155	5.908	5.737
8	11.259	8.649	7.591	7.006	6.632	6.371	6.029	5.814	5.667	5.477	5.359	5.116	4.946
9	10.561	8.022	6.992	6.422	6.057	5.802	5.467	5.257	5.111	4.924	4.808	4.567	4.398
10	10.044	7.559	6.552	5.994	5.636	5.386	5.057	4.849	4.706	4.520	4.405	4.165	3.996
12	9.330	6.927	5.953	5.412	5.064	4.821	4.499	4.296	4.155	3.972	3.858	3.619	3.449
14	8.862	6.515	5.564	5.035	4.695	4.456	4.140	3.939	3.800	3.619	3.505	3.266	3.094
16	8.531	6.226	5.292	4.773	4.437	4.202	3.890	3.691	3.553	3.372	3.259	3.018	2.845
18	8.285	6.013	5.092	4.579	4.248	4.015	3.705	3.508	3.371	3.190	3.077	2.835	2.660
20	8.096	5.849	4.938	4.431	4.103	3.871	3.564	3.368	3.231	3.051	2.938	2.695	2.517
24	7.823	5.614	4.718	4.218	3.895	3.667	3.363	3.168	3.032	2.852	2.738	2.492	2.310
28	7.636	5.453	4.568	4.074	3.754	3.528	3.226	3.032	2.896	2.716	2.602	2.354	2.167
32	7.499	5.336	4.459	3.969	3.652	3.427	3.127	2.934	2.798	2.618	2.503	2.252	2.062
36	7.396	5.248	4.377	3.890	3.574	3.351	3.052	2.859	2.723	2.543	2.428	2.175	1.981
40	7.314	5.179	4.313	3.828	3.514	3.291	2.993	2.801	2.665	2.484	2.369	2.114	1.917
50	7.171	5.057	4.199	3.720	3.408	3.186	2.890	2.698	2.562	2.382	2.265	2.007	1.803
60	7.077	4.977	4.126	3.649	3.339	3.119	2.823	2.632	2.496	2.315	2.198	1.936	1.726
70	7.011	4.922	4.074	3.600	3.291	3.071	2.777	2.585	2.450	2.268	2.150	1.886	1.672
80	6.963	4.881	4.036	3.563	3.255	3.036	2.742	2.551	2.415	2.233	2.115	1.849	1.630
100	6.895	4.824	3.984	3.513	3.206	2.988	2.694	2.503	2.368	2.185	2.067	1.797	1.572

Table 10: F distribution, $1 - p = 0.999$

v_2 \ v_1	1	2	3	4	5	6	8	10	12	16	20	40	120
1	405284	499999	540379	562500	576405	585937	598144	605621	610668	617045	620908	628712	633972
2	998.50	999.00	999.17	999.25	999.30	999.33	999.37	999.40	999.42	999.44	999.45	999.47	999.49
3	167.03	148.50	141.11	137.10	134.58	132.85	130.62	129.25	128.32	127.14	126.42	124.96	123.97
4	74.137	61.246	56.177	53.436	51.712	50.525	48.996	48.053	47.412	46.597	46.100	45.089	44.400
5	47.181	37.122	33.202	31.085	29.752	28.834	27.649	26.917	26.418	25.783	25.395	24.602	24.060
6	35.507	27.000	23.703	21.924	20.803	20.030	19.030	18.411	17.989	17.450	17.120	16.445	15.981
7	29.245	21.689	18.772	17.198	16.206	15.521	14.634	14.083	13.707	13.226	12.932	12.326	11.909
8	25.415	18.494	15.829	14.392	13.485	12.858	12.046	11.540	11.194	10.752	10.480	9.919	9.532
9	22.857	16.387	13.902	12.560	11.714	11.128	10.368	9.894	9.570	9.154	8.898	8.369	8.001
10	21.040	14.905	12.553	11.283	10.481	9.926	9.204	8.754	8.445	8.048	7.804	7.297	6.944
12	18.643	12.974	10.804	9.633	8.892	8.379	7.710	7.292	7.005	6.634	6.405	5.928	5.593
14	17.143	11.779	9.729	8.622	7.922	7.436	6.802	6.404	6.130	5.776	5.557	5.098	4.773
16	16.120	10.971	9.006	7.944	7.272	6.805	6.195	5.812	5.547	5.205	4.992	4.545	4.226
18	15.379	10.390	8.487	7.459	6.808	6.355	5.763	5.390	5.132	4.798	4.590	4.151	3.836
20	14.819	9.953	8.098	7.096	6.461	6.019	5.440	5.075	4.823	4.495	4.290	3.856	3.544
24	14.028	9.339	7.554	6.589	5.977	5.550	4.991	4.638	4.393	4.074	3.873	3.447	3.136
28	13.498	8.931	7.193	6.253	5.656	5.241	4.695	4.349	4.109	3.795	3.598	3.176	2.864
32	13.117	8.639	6.936	6.014	5.429	5.021	4.485	4.145	3.908	3.598	3.403	2.983	2.670
36	12.832	8.420	6.744	5.836	5.260	4.857	4.328	3.992	3.758	3.451	3.258	2.839	2.524
40	12.609	8.251	6.595	5.698	5.128	4.731	4.207	3.874	3.642	3.338	3.145	2.727	2.410
50	12.222	7.956	6.336	5.459	4.901	4.512	3.998	3.671	3.443	3.142	2.951	2.533	2.211
60	11.973	7.768	6.171	5.307	4.757	4.372	3.865	3.541	3.315	3.017	2.827	2.409	2.082
70	11.799	7.637	6.057	5.201	4.656	4.275	3.773	3.452	3.227	2.930	2.741	2.322	1.991
80	11.671	7.540	5.972	5.123	4.582	4.204	3.705	3.386	3.162	2.867	2.677	2.258	1.924
100	11.495	7.408	5.857	5.017	4.482	4.107	3.612	3.296	3.074	2.780	2.591	2.170	1.829

Table 11: Distribution of the Studentized range

$1 - p = 0.95$

ν	2	3	4	5	6	7	a 8	9	10	12	14	16	20
1	17.97	26.98	32.82	37.08	40.41	43.12	45.40	47.36	49.07	51.96	54.32	56.32	59.55
2	6.08	8.33	9.80	10.88	11.73	12.43	13.03	13.54	13.99	14.75	15.37	15.91	16.77
3	4.50	5.91	6.82	7.50	8.04	8.48	8.85	9.18	9.46	9.95	10.35	10.69	11.24
4	3.93	5.04	5.76	6.29	6.71	7.05	7.35	7.60	7.83	8.21	8.52	8.79	9.23
5	3.64	4.60	5.22	5.67	6.03	6.33	6.58	6.80	6.99	7.32	7.60	7.83	8.21
6	3.46	4.34	4.90	5.30	5.63	5.90	6.12	6.32	6.49	6.79	7.03	7.24	7.59
7	3.34	4.16	4.68	5.06	5.36	5.61	5.82	6.00	6.16	6.43	6.66	6.85	7.17
8	3.26	4.04	4.53	4.89	5.17	5.40	5.60	5.77	5.92	6.18	6.39	6.57	6.87
9	3.20	3.95	4.41	4.76	5.02	5.24	5.43	5.59	5.74	5.98	6.19	6.36	6.64
10	3.15	3.88	4.33	4.65	4.91	5.12	5.30	5.46	5.60	5.83	6.03	6.19	6.47
12	3.08	3.77	4.20	4.51	4.75	4.95	5.12	5.26	5.39	5.61	5.80	5.95	6.21
14	3.03	3.70	4.11	4.41	4.64	4.83	4.99	5.13	5.25	5.46	5.64	5.79	6.03
16	3.00	3.65	4.05	4.33	4.56	4.74	4.90	5.03	5.15	5.35	5.52	5.66	5.90
18	2.97	3.61	4.00	4.28	4.49	4.67	4.82	4.96	5.07	5.27	5.43	5.57	5.79
20	2.95	3.58	3.96	4.23	4.45	4.62	4.77	4.90	5.01	5.20	5.36	5.49	5.71
24	2.92	3.53	3.90	4.17	4.37	4.54	4.68	4.81	4.92	5.10	5.25	5.38	5.59
28	2.90	3.50	3.86	4.12	4.32	4.49	4.62	4.74	4.85	5.03	5.18	5.30	5.51
32	2.88	3.48	3.83	4.09	4.28	4.45	4.58	4.70	4.80	4.98	5.12	5.24	5.45
36	2.87	3.46	3.81	4.06	4.25	4.41	4.55	4.66	4.76	4.94	5.08	5.20	5.40
40	2.86	3.44	3.79	4.04	4.23	4.39	4.52	4.63	4.73	4.90	5.04	5.16	5.36
50	2.84	3.42	3.76	4.00	4.19	4.34	4.47	4.58	4.68	4.85	4.98	5.10	5.29
60	2.83	3.40	3.74	3.98	4.16	4.31	4.44	4.55	4.65	4.81	4.94	5.06	5.24
70	2.82	3.39	3.72	3.96	4.14	4.29	4.42	4.53	4.62	4.78	4.91	5.03	5.21
80	2.81	3.38	3.71	3.95	4.13	4.28	4.40	4.51	4.60	4.76	4.89	5.00	5.18
100	2.81	3.36	3.70	3.93	4.11	4.26	4.38	4.48	4.58	4.73	4.86	4.97	5.15

Appendix A

Appendix A: Introduction to matrix algebra

A.1 Some basic definitions

Definition: A vector is an ordered set of numbers. Each number has a given position.

Example: $x = \begin{pmatrix} 5 \\ 3 \\ 8 \end{pmatrix}$ is a column vector with 3 elements.

Example: $\mathbf{y} = \begin{pmatrix} y_1 \\ y_2 \\ \vdots \\ y_n \end{pmatrix}$ is a column vector with n elements.

Definition: A matrix is a two-dimensional (rectangular) ordered set of numbers.

Example: $\mathbf{A} = \begin{pmatrix} 1 & 2 & 4 \\ 1 & 6 & 3 \end{pmatrix}$ is a matrix with two rows and three columns.

Example: $\mathbf{B} = \begin{pmatrix} b_{11} & b_{12} & & b_{1c} \\ b_{21} & b_{22} & & \\ & & \ddots & \\ b_{r1} & b_{r2} & & b_{rc} \end{pmatrix}$ is a matrix with r rows and c columns. The general element of the matrix $\mathbf{B}$ is b_{ij}. The first index denotes row, the second index denotes column.

Vectors are often written using lowercase symbols like $\mathbf{x}$, while matrices are often written using uppercase letters like $\mathbf{A}$. Both matrices and vectors are written in **bold**.

A.2 The dimension of a matrix

Definition: A matrix that has r rows and c columns is said to have dimension $r \times c$.
  Definition: A column matrix with n rows has dimension $n \times 1$.
  Definition: A row matrix with m columns has dimension $1 \times m$.
  Definition: A scalar, i.e. a number, is a matrix that has dimension 1×1.

A.3 The transpose of a matrix

Transposing a matrix means to interchange rows and columns. If $\mathbf{A}$ is a matrix of dimension $r \times c$ then the transpose of $\mathbf{A}$ is a matrix of dimension $c \times r$. The transpose operator is denoted with a prime, $'$, so the transpose of $\mathbf{A}$ is denoted with $\mathbf{A}'$ (Some textbooks indicate a transpose by using the letter T).
  For the elements a'_{ij} of $\mathbf{A}'$ it holds that

$$a'_{ij} = a_{ji}$$

If $\mathbf{x} = \begin{pmatrix} x_1 \\ x_2 \\ \vdots \\ x_n \end{pmatrix}$ is a column vector, then $\mathbf{x}' = \begin{pmatrix} x_1 & x_2 & \cdots & x_n \end{pmatrix}$ is a row vector with n elements.

  Example: The transpose of the matrix

$$\mathbf{A} = \begin{pmatrix} 1 & 2 & 4 \\ 1 & 6 & 3 \end{pmatrix}$$

is

$$\mathbf{A}' = \begin{pmatrix} 1 & 1 \\ 2 & 6 \\ 4 & 3 \end{pmatrix}.$$

A.4 Some special types of matrices

Definition: A matrix where the number of rows = number of columns (i.e. $r = c$) is a square matrix.

Definition: A square matrix that is unchanged when transposed is symmetric.

Example: The matrix $\mathbf{A} = \begin{pmatrix} 3 & 0 & -1 \\ 0 & 1 & 2 \\ -1 & 2 & 4 \end{pmatrix}$ is square and symmetric.

Definition: The elements a_{ii} in a square matrix are called the diagonal elements.

Definition: An identity matrix $\mathbf{I}$ is a symmetric matrix where all elements are 0, except that the diagonal elements are 1: $\mathbf{I} = \begin{pmatrix} 1 & 0 & & 0 \\ 0 & 1 & & 0 \\ & & \ddots & \\ 0 & 0 & & 1 \end{pmatrix}$

Definition: A diagonal matrix is a matrix where all elements are 0, except for the diagonal elements: $\mathbf{D}(a_i) = \begin{pmatrix} a_1 & 0 & & 0 \\ 0 & a_2 & & 0 \\ & & \ddots & \\ 0 & 0 & & a_r \end{pmatrix}$.

Definition: A unit vector is a vector where all elements are 1: $\mathbf{1} = \begin{pmatrix} 1 \\ 1 \\ \vdots \\ 1 \end{pmatrix}$. The transpose is $\mathbf{1}' = \begin{pmatrix} 1 & 1 & \cdots & 1 \end{pmatrix}$.

A.5 Calculations on matrices

Addition, subtraction and multiplication can be defined for matrices.

Definition: Equality: Two matrices A and B with the same dimension $r \times c$ are equal if and only if $a_{ij} = b_{ij}$ for all i and j, i.e. if all elements are equal.

Definition: Addition: The sum of two matrices $\mathbf{A}$ and $\mathbf{B}$ that have the same dimension is the matrix that consists of the sum of the elements of $\mathbf{A}$ and $\mathbf{B}$.

Example: If $\mathbf{A} = \begin{pmatrix} 1 & 2 & 4 \\ 1 & 6 & 3 \end{pmatrix}$ and $\mathbf{B} = \begin{pmatrix} 3 & 9 & 6 \\ 4 & 2 & 1 \end{pmatrix}$ then

$$\mathbf{A} + \mathbf{B} = \begin{pmatrix} 4 & 11 & 10 \\ 5 & 8 & 4 \end{pmatrix}.$$

Example: If $\mathbf{A} = \begin{pmatrix} a_{11} & a_{12} & a_{1c} \\ a_{21} & a_{22} & \\ a_{r1} & a_{r2} & a_{rc} \end{pmatrix}$ and $\mathbf{B} = \begin{pmatrix} b_{11} & b_{12} & b_{1c} \\ b_{21} & b_{22} & \\ b_{r1} & b_{r2} & b_{rc} \end{pmatrix}$ then

$$\mathbf{A} + \mathbf{B} = \begin{pmatrix} a_{11} + b_{11} & a_{12} + b_{12} & a_{1c} + b_{1c} \\ a_{21} + b_{21} & a_{22} + b_{22} & \\ a_{r1} + b_{r1} & a_{r2} + b_{r2} & a_{rc} + b_{rc} \end{pmatrix}.$$

Definition: Subtraction: Matrix subtraction is defined in an analogous way. It holds that

$$\begin{aligned} \mathbf{A} + \mathbf{B} &= \mathbf{B} + \mathbf{A} \\ \mathbf{A} + (\mathbf{B} + \mathbf{C}) &= (\mathbf{A} + \mathbf{B}) + \mathbf{C} \\ \mathbf{A} - (\mathbf{B} - \mathbf{C}) &= (\mathbf{A} - \mathbf{B}) + \mathbf{C} \end{aligned}$$

For matrices that do not have the same dimensions, addition and subtraction are not defined.

A.6 Matrix multiplication

A.6.1 Multiplication by a scalar

To multiply a matrix $\mathbf{A}$ by a scalar (= a number) c means that all elements in $\mathbf{A}$ are multiplied by c.

Example: If $\mathbf{A} = \begin{pmatrix} 1 & 2 & 4 \\ 1 & 6 & 3 \end{pmatrix}$ then $4 \cdot \mathbf{A} = \begin{pmatrix} 4 & 8 & 16 \\ 4 & 24 & 12 \end{pmatrix}$

Example: $k \cdot \mathbf{A} = \begin{pmatrix} k \cdot a_{11} & k \cdot a_{12} & k \cdot a_{1c} \\ k \cdot a_{21} & k \cdot a_{22} & \\ k \cdot a_{r1} & k \cdot a_{r2} & k \cdot a_{rc} \end{pmatrix}.$

A.6.2 Multiplication by a matrix

Matrix multiplication of type $\mathbf{C} = \mathbf{A} \cdot \mathbf{B}$ is defined only if the number of columns in $\mathbf{A}$ is equal to the number of rows in $\mathbf{B}$. If $\mathbf{A}$ has dimension $p \times r$ and $\mathbf{B}$ has dimension $r \times q$ then the product $\mathbf{A} \cdot \mathbf{B}$ will have dimension $p \times q$. The elements of $\mathbf{C}$ are calculated as

$$c_{ij} = \sum_{k=1}^{r} a_{ik} b_{kj}.$$

Example: If $\mathbf{A} = \begin{pmatrix} 1 & 2 & 3 \\ -1 & 0 & 1 \end{pmatrix}$ and $\mathbf{B} = \begin{pmatrix} 6 & 5 & 4 \\ -1 & 1 & -1 \\ 0 & 2 & 0 \end{pmatrix}$ then

$$\mathbf{A} \cdot \mathbf{B} =$$
$$\begin{pmatrix} 1 \cdot 6 - 2 \cdot 1 + 3 \cdot 0 & 1 \cdot 5 + 2 \cdot 1 + 3 \cdot 2 & 1 \cdot 4 - 2 \cdot 1 + 3 \cdot 0 \\ -1 \cdot 6 - 0 \cdot 1 + 1 \cdot 0 & -1 \cdot 5 + 0 \cdot 1 + 1 \cdot 2 & -1 \cdot 4 - 0 \cdot 1 + 1 \cdot 0 \end{pmatrix}$$
$$= \begin{pmatrix} 4 & 13 & 2 \\ -6 & -3 & -4 \end{pmatrix}.$$

A.6.3 Calculation rules of multiplication

It holds that

$$\mathbf{A}(\mathbf{B} + \mathbf{C}) = \mathbf{A} \cdot \mathbf{B} + \mathbf{A} \cdot \mathbf{C}$$
$$\mathbf{A}(\mathbf{B} \cdot \mathbf{C}) = (\mathbf{A} \cdot \mathbf{B}) \cdot \mathbf{C}.$$

Note that in general, $\mathbf{AB} \neq \mathbf{BA}$. The order has importance for multiplication. In the expression $\mathbf{AB}$ the matrix $\mathbf{A}$ has been post-multiplied with the matrix $\mathbf{B}$. In the expression $\mathbf{BA}$ the matrix $\mathbf{A}$ has been pre-multiplied with the matrix $\mathbf{B}$. Note that $(\mathbf{AB})' = \mathbf{B}'\mathbf{A}'$.

A.6.4 Idempotent matrices

Definition: A matrix $\mathbf{A}$ is idempotent if $\mathbf{A} \cdot \mathbf{A} = \mathbf{A}$.

A.7 The inverse of a matrix

Definition: The inverse of a square matrix $\mathbf{A}$ is the unique matrix $\mathbf{A}^{-1}$ for which it holds that $\mathbf{A}\mathbf{A}^{-1} = \mathbf{A}^{-1}\mathbf{A} = \mathbf{I}$. That is: the matrix multiplied with its inverse results in the unit matrix. (Note that the same rule holds for scalars: $3 \cdot 3^{-1} = 3 \cdot \frac{1}{3} = \frac{3}{3} = 1$).

Example: The inverse of the matrix $\mathbf{A} = \begin{pmatrix} 5 & 10 \\ 3 & 2 \end{pmatrix}$ is

$$\mathbf{A}^{-1} = \begin{pmatrix} -0.1 & 0.5 \\ 0.15 & -0.25 \end{pmatrix}.$$

To verify this we calculate

$$\mathbf{A} \cdot \mathbf{A}^{-1} = \begin{pmatrix} 5 & 10 \\ 3 & 2 \end{pmatrix} \begin{pmatrix} -0.1 & 0.5 \\ 0.15 & -0.25 \end{pmatrix}$$

$$= \begin{pmatrix} 5 \cdot (-0.1) + 10 \cdot 0.15 & 5 \cdot 0.5 + 10 \cdot (-0.25) \\ 3 \cdot (-0.1) + 2 \cdot 0.15 & 3 \cdot 0.5 + 2 \cdot (-0.25) \end{pmatrix}$$

$$= \begin{pmatrix} 1.0 & 0 \\ 0 & 1.0 \end{pmatrix} = \mathbf{I}.$$

It is possible that the inverse $\mathbf{A}^{-1}$ does not exist. $\mathbf{A}$ is then said to be singular. The following relations hold for inverses:
The inverse of a symmetric matrix is symmetric

$$(\mathbf{A}')^{-1} = (\mathbf{A}^{-1})'.$$

The inverse of a product of several matrices is obtained by taking the product of the inverses, in opposite order:

$$(\mathbf{ABC})^{-1} = \mathbf{C}^{-1}\mathbf{B}^{-1}\mathbf{A}^{-1}.$$

If c is a scalar different from zero, then

$$(c\mathbf{A})^{-1} = \frac{1}{c}\mathbf{A}^{-1}.$$

A.8 Generalized inverses

A matrix $\mathbf{B}$ is said to be a generalized inverse of the matrix $\mathbf{A}$ if $\mathbf{ABA} = \mathbf{A}$. The generalized inverse of a matrix $\mathbf{A}$ is denoted with $\mathbf{A}^-$. If $\mathbf{A}$ is nonsingular then $\mathbf{A}^- = \mathbf{A}^{-1}$. When $\mathbf{A}$ is singular, $\mathbf{A}^-$ is not unique. A generalized inverse of a matrix $\mathbf{A}$ can be calculated as

$$\mathbf{A}^- = (\mathbf{A}'\mathbf{A})^{-1}\mathbf{A}'.$$

A.9 The rank of a matrix

Definition: Two vectors are linearly dependent if the elements of one vector are proportional to the elements of the other vector.
   Example: If $\mathbf{x}' = (\; 1 \;\; 0 \;\; 1 \;)$ and $\mathbf{y}' = (\; 4 \;\; 0 \;\; 4 \;)$ then the vectors $\mathbf{x}$ and $\mathbf{y}$ are linearly dependent.
   Definition: A set of vectors are linearly independent if it is impossible to write any one of the vectors as a linear combination of the others.

Example: The vectors $\mathbf{t}' = \begin{pmatrix} 1 & 0 & 0 \end{pmatrix}$, $\mathbf{u}' = \begin{pmatrix} 0 & 1 & 0 \end{pmatrix}$ and $\mathbf{v}' = \begin{pmatrix} 0 & 0 & 1 \end{pmatrix}$ are linearly independent.

Definition: The degree of linear independence among a set of vectors is called the rank of the matrix that is composed by the vectors.

The following properties hold for the rank of a matrix:

The rank of $\mathbf{A}^{-1}$ is equal to the rank of $\mathbf{A}$.

The rank of $\mathbf{A}'\mathbf{A}$ is equal to the rank of $\mathbf{A}$ (It is also true that the rank of $\mathbf{A}\mathbf{A}'$ is equal to the rank of $\mathbf{A}$).

The rank of a matrix $\mathbf{A}$ does not change if $\mathbf{A}$ is pre- or postmultiplied with a nonsingular matrix.

A.10 Determinants

To each square matrix $\mathbf{A}$ belongs a unique scalar that is called the determinant of A. The determinant of A is written as $|\mathbf{A}|$. The determinant of a matrix of dimension n can be calculated as $|\mathbf{A}| = \sum (-1)^{\#(\pi(n))} \prod_{i=1}^{n} a_{\pi_i, i}$.

Here, $\pi(n)$ denotes any permutation of the numbers $1, 2, \ldots n$. $\#\pi(n)$ denotes the number of inversions of a permutation $\pi(n)$. This is the number of exchanges of pairs of the numbers in $\pi(n)$ that are needed to bring them back into natural order. Determinants of small matrices can be calculated by hand, but for larger matrices we prefer to leave the work to computers.

If $\mathbf{A}$ is singular, then the determinant $|\mathbf{A}| = 0$.

A.11 Eigenvalues and eigenvectors

To each symmetric square matrix $\mathbf{A}$ of dimension $n \times n$ belongs n scalars λ_i that are called the eigenvalues of $\mathbf{A}$. These are solutions to the equation

$$|\mathbf{A} - \lambda \mathbf{I}| = 0.$$

The eigenvalues have the following properties:

The product of all eigenvalues of $\mathbf{A}$ is equal to $|\mathbf{A}|$.

The sum of all eigenvalues of $\mathbf{A}$ is equal to $tr(\mathbf{A})$, which is the sum of the diagonal elements of $\mathbf{A}$. The symbol $tr(\mathbf{A})$ can be read as "the trace of $\mathbf{A}$".

To each eigenvalue λ_i belongs a vector $\mathbf{v}_i$ for which the relation

$$\mathbf{A}\mathbf{v}_i = \lambda_i \mathbf{v}_i$$

holds. The vectors $\mathbf{v}_i$ is called eigenvectors to $\mathbf{A}$.

A.12 Some statistical formulas on matrix form

$$\mathbf{x}'\mathbf{x} = \begin{pmatrix} x_1 & x_2 & \ldots & x_n \end{pmatrix} \begin{pmatrix} x_1 \\ x_2 \\ \vdots \\ x_n \end{pmatrix} = \sum_{i=1}^{n} x_i^2$$

$$\mathbf{x}'\mathbf{y} = \begin{pmatrix} x_1 & x_2 & \ldots & x_n \end{pmatrix} \begin{pmatrix} y_1 \\ y_2 \\ \vdots \\ y_n \end{pmatrix} = \sum_{i=1}^{n} x_i y_i$$

$$\mathbf{1}'\mathbf{y} = \sum_{i=1}^{n} y_i \qquad \mathbf{1}'\mathbf{1} = n \qquad \mathbf{1}'\mathbf{y} n^{-1} = (\mathbf{1}'\mathbf{1})^{-1} \mathbf{1}'\mathbf{y} = \overline{y}$$

A.13 Further reading

This chapter has only given a very brief and sketchy introduction to matrix algebra. A more complete treatment can be found in textbooks such as Searle (1982).

Bibliography

[1] Aird, I., Benthall, H. H., Mehigan, J. A. and Robers, J. A. F. (1954): The blood groups in relation to peptic ulceration and carcinoma of colon, rectum, breast and bronchus: an association between the ABO blood groups and peptic ulceration. *British Medical Journal,* ii, 315-321.

[2] Anscombe, F. J. (1973): Graphs in Statistical Analysis. *The American Statistician,*. **27**, 17-21.

[3] Armitage, P. and Colton, T. (1998): *Encyclopedia of Biostatistics.* Chichester, Wiley.

[4] Barniot, N. A. and Brothwell, D. R. (1959): The evaluation of metrical data in the comparison of modern and ancient bones. In: Wolstenholme and O'Connor (Eds.): Medical biology and etruscan origins. Little Brown and Co. New York.

[5] Betz, H. -D., Kulzer, R., König, H. L. Tritschler J and Wagner H. (1990): Dowsing reviewed — the effect persists. *Naturwissenschaften,* **83**, 272-275.

[6] Blackwelder, W. C. (1982): "Proving the null hypothesis" in clinical trials. *Controlled clinical trials,* 3:345-353.

[7] Bortkewitsch, L. von (1898). *Das Gesetz der kleinen Zahlen.* Leipzig: G. Teubner.

[8] Box, G. E. P. (1950): Promlems in the analysis of growth and wear curves. *Biometrics,* **6**, 362-389.

[9] Box, G. E. P. and Cox, D. R. (1964): An analysis of transformations. *JRSS B,* **26**, 211-246.

[10] Carroll, R. J. and Ruppert, D. (1984): Power Transformations when Fitting Theoretical Models to Data, *Journal of the American Statistical Association*, **79**, 321-328.

[11] Casagrande, J. T. and Pike, M. C. (1978): An improved approximate formula for calculating sample sizes for comparing two binomial distributions. *Biometrics*, 34, 483-486.

[12] Christensen, R. (1996): Analysis of variance, design and regression. London, Chapman & Hall.

[13] Cicirelli, M. F., Robinson, K. R. and Smith, L. D. (1983): Internal pH of Xenopus oocytes: a study of the mechanism and role of pH changes during meiotic maturation. *Developmental biology*, 100, 133-146.

[14] Clarke, R. D. (1946): An application of the Poisson distribution. *J. of the Inst. of Actuaries*.

[15] Collett, D. (1991): *Modelling binary data*. London, Chapman & Hall.1

[16] Connett, J. E., Smith, J. A. and McHugh, R. B. (1987): Sample size and power for pair-matched case-control studies. *Statistics in Medicine*, 6, 53-59.

[17] Day, S. (1999): *Dictionary for clinical trials*. New York, Wiley.

[18] Dale et al (1987): Beta endorphin: a factor in "fun run" collapse? *British Medical Journal*, 294, 1004.

[19] Darwin, Charles (1876): *The effects of cross- and self-fertilization in the vegetable kingdom*. London, John Murray.

[20] DeMets, D. L. and Lan, K. K. G. (1994): Interim analysis: the alpha spending function approach. *Statistics in Medicine*, 13, 1341-1352.

[21] Didon, U. and Olsson, U. (1997): A Bioassay method for detection of chlorinated phenoxy acids in water. *Swedish J. Agr. Res. 27*, 121-128.

[22] Dietl, J. (1849): *Der Aderlass in der Lungenentzündung*.

[23] Digby, P. G. N. and Kempton, R. A. (1987): *Multivariate analysis of ecological communities*. London, Chapman & Hall.

[24] Doll, R. and Hill, A. B. (1954): The Mortality of Doctors in Relation to Their Smoking Habits. British Medical Journal p 1451.

[25] Draper, N. R. and Stoneman, D.M. (1966): Testing for the Inclusion of Variables in Linear Regression by a Randomisation Technique. *Technometrics*, **8**, 695-699

[26] Draper, N. and Smith, H. (1998): *Applied regression analysis*, 2nd ed. New York, Wiley.

[27] Dunnett, C. W. (1955): A multiple comparison procedure for comparing several treatments with a control. *J. Am. Stat. Assoc*, **50**, 1096-1121.

[28] Engstrand, U. och Olsson, U. (2003): *Variansanalys och försöksplanering*. Lund, Studentlitteratur.

[29] Finney, D. J. (1947, 1952): *Probit analysis. A statistical treatment of the sigmoid response curve.* Cambridge, Cambridge University Press.

[30] Fisher, R. A. (1925): *Statistical methods for research workers*, 14th ed, 1970. New York, Hafner press.

[31] Fleiss, J. L. (1986): *Design and analysis of clinical experiments.* New York, Wiley.

[32] Fleming, T. R. (1982): One-sample multiple testing procedure for phase II clinical trials. *Biometrics*, 38, 143-151.

[33] Friedman, L. M., Furberg, C. D. and DeMets, D. L. (1998): Fundamentals of clinical trials. New York, Springer.

[34] Geller, N. L. and Pocock, S. J. (1987): Interim analyses in randomized clinical trials: ramifications and guidelines for practitioners. *Biometrics*, 43, 213-223.

[35] Gooley, T. A., Martin, P. J., Fisher, L. D. and Pettinger, M. (1994): Simulation as a design tool for phase I/II clinical trials: and example from bone marrow transplantation. *Controlled clinical trials*, 15, 450-462.

[36] Gosset, W. S. ("Student") (1908): The probable error of a mean. *Biometrika*, Vol. 6, No. 1. pp. 1-25.

[37] Gäredal, L. (1998): Greenhouse cultivation of tomatoes (Lycopersicon esculentum Mill.) in limited growing beds, based on nutrients from locally produced farm yard manure compost and fresh green material. SLU, Uppsala, Licentiate thesis.

[38] Haberman, S. (1978): *Analysis of qualitative data.* Vol. 1: Introductory topics. New York, Academic Press.

[39] Hand, D. J., Daly, F., Lunn, A. D., McConway, K. J. och Ostrowski, E. (1994): *A handbook of small data sets.* London, Chapman & Hall.

[40] Heggestad, H. E. and Bennett, J. H. (1981): Photochemical oxidants potentiate yield losses in snap beans attributable to sulfur dioxide. *Science,* **213**, 1008-1010.

[41] Holmberg, L. and Baum, M. (1995): Can results from clinical trials be generalized? *Nature Medicine*, 1, 734-736.

[42] Hróbjartsson, A. (1996): The uncontrollable placebo effect. *Eur J Clin Pharmacol*, 50, 345-348.

[43] Hurn, M. W., Barker, N. W. and Magath, T. D. (1945): The determination of prothrombin time following the administration of dicumarol with specific reference to thromboplastin. *J. Lab. Clin. Med.,* **30**, 432-447.

[44] Jolicoeur, P. and Mosimann, J.E. (1960): Size and shape variation on the painted turtle: a principal component analysis. *Growth,* 24, 339-354.

[45] Jones, K.L., Smith, D.W., Streissguth A.P., och Myrianthopoulos N.C. (1974): Outcome in offspring of chronic alcoholic women. *Lancet*, 1974, Jun 1; 1(7866), 1076-8.

[46] Jones, B. and Kenward, M. G. (1989): *Design and analysis of crossover trials.* London, Chapman and Hall.

[47] Kettlewell, H. B. D. (1955a): How industrialisation can alter species. *Discovery* 16 (12): 507-11.

[48] Kettlewell, H. B. D. (1955b): Selection experiments in Industrial Melanism in the Lepidoptera. *Heredity* 9 (3): 323-42.

[49] Klein, J. and Moeschberger, M. (1997): *Survival analysis: techniques for censored and truncated data.* New York, Springer.

[50] Koch, G. G. and Edwards, S. (1988): Clinical efficiacy trials with categorical data. In: *Biopharmaceutical statistics for drug development*, K. E. Peace, ed. New York, Marcel Dekker, pp. 403-451.

[51] Kohlmeier L, Arminger G, Bartolomeycik S, Bellach B, Rehm J, Thamm M. (1992): Pet birds as an independent risk factor for lung cancer: case-control study. *British Medical Journal*, pp 986-989

[52] Korn, E. L., Yu, K. F. and Miller, L. L. (1993): Stopping a clinical trial very early because of toxicity: summarizing the evidence. *Controlled clinical trials*, 14, 286-295.

[53] Kupper, L. L. and Hafner, K. B. (1989): How appropriate are popular sample size formulas? *The American statistician*, 43, 2, 101-105.

[54] Lachin, J. M. and Foulkes, M. A. (1986): Evaluation of sample size and power for analyses of survival with allowance for nonuniform patient entry, losses to follow-up and stratification. *Biometrics*, 42, 507-519.

[55] Lakatos, E. (1986): Sample size determination in clinical trials with time-dependent rates of losses and noncompliance. *Controlled clinical trials*, 7, 189-199.

[56] Lakatos, E. (1988): Sample sizes based on the log-rank statistic in complex clinical trials. *Biometrics*, 44, 229-241.

[57] Lawal, R.O. (1992): The effect of a single oral dose of polyphenols obtained from the outercoat of the fruit of Treculia africana in protein-deficient rats. *Food Chemistry*, **44** 321-323.

[58] Lefcoe, N M and Wonnacott, T H : The prevalence of chronic respiratory disease in four occupational groups. *Archives of Environmental Health*, 29, 143-146.

[59] Lilja, C. (1987): A note on the metabolic rate in the young rat. *Swedish J. agric. Res.* **17**, 103-104.

[60] Lilja, C. and Olsson, U: Changes in embryonic development associated with long-term selection for high growth rate in Japanese Quail. *Growth*, 1987, **51**, 301-308.

[61] Lindahl, B., Stenlid, J., Olsson, S. and Finlay, R. (1999): Translocation of ^{32}P between interacting mycelia of a wood-decomposing fungus and ectomycorrhizal fungi in microcosm systems. *New Phytol.*, **144**, 183-193.

[62] Linde, K., Clausius, N., Ramirez, G., Meichart, D., Eitel, F., Hedges, L. V and Jonas, W. B. (1997): Are the clinical effects of homoeopathy placebo effects? A meta-analysis of placebo-controlled trials. *The Lancet*, 350, 834-843.

[63] Liss, P., Nygren, A., Olsson, U., Ulfendahl, H. R. and Eriksson, U.: Effects of contrast media and Mannitol on renal medullary blood flow and renal blood cell aggregation in the rat kidney. *Kidney International*, 1996, **49**, 1268-1275.

[64] Medical Research Council (1948): Streptomycin treatment of pulmonary tuberculosis. *British Medical Journal*, **ii**, 769-782.

[65] Medical Research Council (1950): Clinical trials of antihistamine drugs in the prevention and treatment of the common cold. *British Medical Journal*, **ii**, 425-429.

[66] Minitab Inc. (2004): *Minitab reference manual, release 14*. Minitab Inc., State College, PA.

[67] Nelder, J. A. (1971), Contribution to the Discussion of Paper by R. O'Neill and G. B. Wetherill Journal of the Royal Statistical Society, Series B, 33, 244-246.

[68] Noether, G. E. (1987): Sample size determination for some common nonparametric tests. *JASA*, 82, 398, 645-647.

[69] Norton, P.G. & Dunn, E.V. (1985): Snoring as a risk factor for disease. *Brit. Med. J.* pp 630-632.

[70] O'Brien, P. C. (1984): Procedures for comparing samples with multiple endpoints. *Biometrics*, 40, 1079-1087.

[71] Olsson, U. (2002): *Generalized linear models – an applied approach*. Lund, Studentlitteratur.

[72] Olsson, U., Englund, J-E. and Engstrand, U. (2005): *Biometri; Grundläggande biologisk statistik*. Lund, Studentlitteratur.

[73] Ostle, B. (1963): *Statistics in research*. Ames, Iowa: Iowa state university press.

[74] Owen, D. F. (1963): Polymorphism and Population Density in the African Land Snail,. *Science*, 140, 3567, pp. 666-667.

[75] Owen, D. B. (1965): A special case of a bivariate non-central t distribution. *Biometrika*, 52, 3 and 4, p. 437-446.

[76] Patel, H. I. (1992): Sample size for a dose-response study. *Journal of biopharmaceutical statistics*, 2, 1-8.

[77] Piantadosi, S. (1997): *Clinical trials: a methodological perspective.* New York, Wiley.

[78] Pignon, J-P., Tarayre, M., Auquier, A., Arriagada, R., le Chevalier, T., Ruffié, P., Rivière, A., Monnet, I., Chomy, P. and Tuchais, C. (1994): Triangular test and randomized trials: practical problems in a small cell lung cancer trial.

[79] Pocock, S. J. (1983): *Clinical trials. A practical approach.* New York, Wiley.

[80] Pocock, S. J. (1977): Group sequential methods in the design and analysis of clinical trials. *Biometrika*, 64, 2, 191-199.

[81] Pocock, S. J. (1982): Interim analyses for randomized clinical trials: the group sequential approach. *Biometrics*, 38, 153-162.

[82] Rea, T. M., Nash, J. F., Zabik, J. E., Born, G. S. och Kessler, W. V. (1984): Effects of Toulene inhalation on brain biogenic amines in the rat. *Toxicology*, **31**, 143-150.

[83] Rodary, C., Com-Nougue, C. and Tournade, M-F. (1989): How to establish equivalence between treatments: a one-sided clinical trial in paedriatic oncology. *Statistics in Medicine*, 8, 593-598.

[84] SAS Institute Inc. (2008): *SAS/STAT 9.2 User's Guide.*, SAS Institute, Cary, NC.

[85] Samuels, M. and Witmer, J. A. (1999): *Statistics for the life sciences.* Upper Saddle River, NJ: Prentice-Hall.

[86] Schwarz, C. J. (2005): Data archive posted at http://www.stat.sfu.ca/~cschwarz/Stat-650/Notes/MyPrograms/

[87] Sen, A.K. and Srivastava, M. S. (1990): *Regression analysis: theory, methods and applications.* New York, Springer.

[88] Senn, S. (1997): *Statistical issues in drug development.* New York, Wiley.

[89] Shih, J. H. (1995): Sample size calculations for complex clinical trials with survival endpoints. *Controlled clinical trials*, 16, 395-407.

[90] Sillanaukee, P. and Olsson, U. (2001): Improved diagnostic classification of alcohol abusers by combining carbohydrate deficient transferrin and gamma-glutamyltransferate. Clinical Chemistry, 2001, 47, 681.

[91] Simon, R. (1989): Optimal two-stage designs for phase II clinical trials. *Controlled clinical trials*, 10, 1-10.

[92] Snedecor, G. W. and Cochran, W. G. (1980): *Statistical methods*. Ames, Iowa: Iowa State University Press.

[93] Sokal, R. R. och Karten, M. (1964): Competition among genotypes of Tribolium castaneum at varying densities and gene frequencies. *Genetics*, 49, 195-211.

[94] Sokal, R. R. och Rohlf, F. J. (1973): *Introduction to biostatistics*. San Fransisco, Freeman.

[95] Sokal, R. R. och Rohlf, F. J. (1981): *Biometry*, 2nd edition. San Fransisco, Freeman.

[96] Storer, B. (1989): Design and analysis of Phase I clinical trials. Biometrics, 49, 1117-1125.

[97] Student (1908), The Probable Error of a Mean *Biometrika*, **6**, 1–25.

[98] Tripepi, R. R. and Mitchell, C. A. (1984): Metabolic Response of River Birch and European Birch and European Birch Roots to Hypoxia . Plant Physiology, 76, 31-35.

[99] Vejde, Olle och Leander, Eva: *Ordbok i statistik*. Borlänge, Olle Vejde förlag.

[100] Weindling, A. M., Bamford, F. N. och Whittall, R. A. (1986): Health of juvenile delinquents. *British Medical Journal*, **292**, 447.

[101] Whitehead, J. (1993): Sample size calculations for ordered categorical data. *Statistics in Medicine*, 12, 2257-2271.

[102] Yates, F (1934). Contingency table involving small numbers and the χ^2 test. Supplement to the Journal of the Royal Statistical Society 1(2): 217–235.

[103] Zagal, E., Bjarnason, S. and Olsson, U. (1993): Carbon and nitrogen in the root-zone of Barley supplied with nitrogen fertilizer at two rates. *Plant and Soil*, **157**, 51-63.

[104] Åman, P. (1988): The variation in chemical composition of Swedish wheats. *Swedish J. Agric. Res.* **18** 27-30.

Answers to the exercises

1.1 A. A Minitab printout gives the equation as:

Regression Analysis: Yield versus SO2

```
The regression equation is
Yield = 1.36 - 2.08 SO2
```

B.
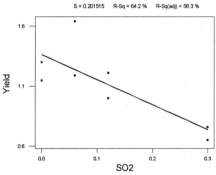

C. Analysis of Variance

```
Source              DF        SS         MS        F         P
Regression           1     0.43750    0.43750    10.77     0.017
Residual Error       6     0.24365    0.04061
Total                7     0.68115
```

The residual variance is estimated as $\widehat{\sigma}_e^2 = MS_e = 0.04061$.

D. $H_0: \beta = 0 \quad H_1: \beta \neq 0 \quad F = 10.77$ on (1;6) d.f., $p = 0.017$.
The slope is significantly different from zero at the 5% level.

E. $\widehat{y} = 1.36 - 2.08 \cdot 0.20 = 0.944$. $\overline{x} = 0.12$; $\sum (x - \overline{x})^2 = 0.1008$.
$Var(\widehat{y}) = MS_e \left(1 + \frac{1}{n} + \frac{(x_0 - \overline{x})^2}{\sum (x - \overline{x})^2}\right) = 0.04061 \left(1 + \frac{1}{8} + \frac{(0.20 - 0.12)^2}{0.1008}\right) =$

0.04826 5. The confidence interval is: $0.944 \pm 2.447\sqrt{0.048265}$; 0.944 ± 0.538.

1.2 $b_1 = \frac{\sum(x-\bar{x})(y-\bar{y})}{\sum(x-\bar{x})^2} = \frac{-276}{660} = -0.418$

$b_0 = \bar{y} - b_1\bar{x} = 234.4 + 0.418 \cdot 63 = 260.73$ so the equation of the line is $\widehat{y} = 260.73 - 0.418x$.

$SS_R = b_1 \sum (x-\bar{x})(y-\bar{y}) = -0.418 \cdot (-276) = 115.37$

$SS_T = \sum(y-\bar{y})^2 = 131.2$ gives $SS_e = SS_T - SS_R = 131.2 - 115.37 = 15.83$.

ANOVA table:

Source	d.f.	SS	MS
Regression	1	115.37	115.37
Residual	18	15.83	0.88
Total	19	131.20	

$H_0: \beta_1 = 0$ is tested using $F_{1,18} = MS_x/MS_e = 115.37/0.88 = 131.1$ which is significant ($p < 0.001$).

$R^2 = \frac{SS_R}{SS_T} = \frac{115.37}{131.20} = 0.88$ i.e. $R^2 = 88\%$.

β_0 can here only be interpreted as an intercept; it has no real interpretation.

β_1 is the average improvement per year of the mile-record during this period.

I would hesitate to extrapolate the line to today, in order to forecast the record. We cannot be sure that the record will continue to improve by 0.42 seconds each year: what would that lead to, in 100 years...

1.3 $b_1 = \frac{\sum(x-\bar{x})(y-\bar{y})}{\sum(x-\bar{x})^2} = \frac{32393}{1171} = 27.663$

$b_0 = \bar{y} - b_1\bar{x} = 119.1 - 27.663 \cdot 3.5 = 22.28$ so the equation of the line is $\widehat{y} = 22.28 + 27.663x$.

$SS_R = b_1 \sum (x-\bar{x})(y-\bar{y}) = 27.663 \cdot 32393 = 896088$

$SS_T = \sum(y-\bar{y})^2 = 943055$ gives $SS_e = SS_T - SS_R = 943055 - 896088 = 46967$

ANOVA table:

Source	d.f.	SS	MS
Regression	1	896088	896088.0
Residual	104	46967	451.6
Total	105	943055	

$H_0: \beta_1 = 0$ is tested using $F_{1,104} = MS_x/MS_e = 896088/451.6 = 1984.3$ which is significant ($p < 0.001$).

$R^2 = \frac{SS_R}{SS_T} = \frac{896088}{943055} = 0.95$ i.e. $R^2 = 95\%$.

β_0 is the intercept of the line. β_1 is the change in y for a change in x of one unit.

A 95% confidence interval for β_1: $27.663 \pm 1.984\sqrt{\frac{451.6}{1171}}$, 27.663 ± 1.232.

A 95% confidence interval for a new observation with $x = 10$ is:
$22.28 + 27.663 \cdot 10 \pm 1.984 \sqrt{451.6\left(1 + \frac{1}{106} + \frac{(10-3.5)^2}{1171}\right)}$; 298.91 ± 43.11.

1.4 $b_1 = \frac{\sum(x-\bar{x})(y-\bar{y})}{\sum(x-\bar{x})^2} = \frac{6.52}{3.27} = 1.99$.
$b_0 = \bar{y} - b_1\bar{x} = 5.4 - 1.99 \cdot 2.05 = 1.32$ so the equation of the line is $\hat{y} = 1.32 + 1.99x$.
$SS_R = b_1 \sum (x-\bar{x})(y-\bar{y}) = 1.99 \cdot 6.52 = 12.975$
$SS_T = \sum(y-\bar{y})^2 = 17.45$ gives $SS_e = SS_T - SS_R = 17.45 - 12.975 = 4.475$

ANOVA table:

Source	d.f.	SS	MS
Regression	1	12.975	12.975
Residual	26	4.475	0.172
Total	27	17.450	

$H_0: \beta_1 = 0$ is tested using $F_{1,18} = MS_x/MS_e = 12.975/0.172 = 75.436$ which is significant ($p < 0.001$).
$R^2 = \frac{SS_R}{SS_T} = \frac{12.975}{17.450} = 0.743$ i.e. $R^2 = 74.3\%$ which gives $r = \sqrt{R^2} = \sqrt{0.743} = 0.862$.

β_0 is the intercept of the line. β_1 is the change in the width of the shell for a change in shell height of one unit.

1.5 A Minitab printout gives most of the answers:

Regression Analysis

```
The regression equation is
Test = 3.96 + 7.48 Timmar

Predictor      Coef       StDev          T         P
Constant       3.962      3.159       1.25     0.236
Timmar         7.4780     0.3980     18.79     0.000

S = 5.370      R-Sq = 97.0%      R-Sq(adj) = 96.7%

Analysis of Variance

Source           DF         SS         MS         F         P
Regression        1      10178      10178    352.96     0.000
Residual Error   11        317         29
Total            12      10495
```

A residual plot is:

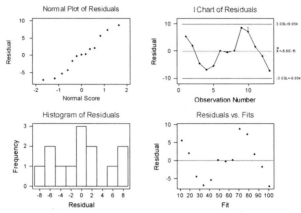

The graph suggests that the true relation might be nonlinear. You can see some systematic pattern in the "I chart".

Minitab can also calculate a confidence interval (CI), as well as a prediction interval (PI):

```
Predicted Values

    Fit  StDev Fit       95.0% CI            95.0% PI
  78.74       1.91  ( 74.54;   82.94)  ( 66.20;   91.29)
```

1.6 A. We get

$$\widehat{\beta}_1 = \frac{\sum (x_i - \overline{x})(y_i - \overline{y})}{\sum (x_i - \overline{x})^2} = \frac{81.90}{2800} = 0.02925,$$

$$\widehat{\beta}_0 = \overline{y} - \widehat{\beta}_1 \overline{x} = 0.83 - 0.02925 \cdot 30 = -0.0475$$

so the regression line is estimated as

$$\widehat{y} = -0.0475 + 0.02925x.$$

B. The ANOVA table is

Source	SS	d.f.	MS
Regression	2.3956	1	2.3956
Residual	0.0352	5	0.0070
Total	2.4308	6	

C. $H_0: \beta_1 = 0$ is tested against $H_1: \beta_1 \neq 0$ using $F = \frac{MS_R}{MS_e} = \frac{2.3956}{0.0070} = 342.23$ on (1,5) d.f.. This is highly significant; the 0.1% limit is 47.181.

D. An observation with $x = 45$ would be predicted to have $\hat{y} = -0.0475 + 0.02925 \cdot 45 = 1.2688$. The prediction interval for this observation is obtained as

$$\hat{y} \pm t_{0.975, 5} \sqrt{MS_e \left(1 + \frac{1}{n} + \frac{(x - \bar{x})^2}{\sum (x_i - \bar{x})^2}\right)}$$

$$1.2688 \pm 2.571 \sqrt{0.0070 \left(1 + \frac{1}{7} + \frac{(45 - 30)^2}{2800}\right)}$$

$$1.2688 \pm 0.2379$$

so the limits are $(1.031 \text{—} 1.507)$.

1.7 A. Model: $y = \beta_0 + \beta_1 x + e$. The parameters are estimated as

$$\hat{\beta}_1 = \frac{\sum xy - \frac{\sum x \sum y}{n}}{\sum x^2 - \frac{(\sum x)^2}{n}} = \frac{31447 - \frac{450 \cdot 649}{10}}{28406 - \frac{450^2}{10}} = 0.27489$$

$$\hat{\beta}_0 = \bar{y} - \hat{\beta}_1 \bar{x} = 64.9 - 0.27489 \cdot 45 = 52.530$$

B. Minitab supplies the following graph:

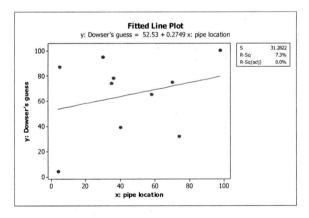

C.
$$r = \frac{\sum xy - \frac{\sum x \sum y}{n}}{\sqrt{\left(\sum x^2 - \frac{(\sum x)^2}{n}\right)\left(\sum y^2 - \frac{(\sum y)^2}{n}\right)}}$$

$$= \frac{31447 - \frac{450 \cdot 649}{10}}{\sqrt{\left(28406 - \frac{450^2}{10}\right)\left(50565 - \frac{649^2}{10}\right)}} = 0.27015$$

D. We have that $SS_T = \sum (y_i - \bar{y})^2 = 50565 - \frac{649^2}{10} = 8444.90$. Furthermore, $SS_R = b_1 SP_{xy} = 0.27489 \left(31447 - \frac{450 \cdot 649}{10}\right) = 616.30$. This gives $SS_e = SS_T - SS_R = 8444.90 - 616.30 = 7828.60$ on $n - 2 = 10 - 2 = 8$ degrees of freedom. ANOVA table:

Source	SS	d.f.	MS
Regression	616.3	1	616.3
Residual	7828.6	8	978.6
Total	8444.9	9	

The hypothesis $H_0\colon \beta_1 = 0$ is tested using $F = \frac{MS_R}{MS_e} = \frac{616.30}{\frac{7828.60}{8}} = 0.63$ on $(1,8)$ d.f.. The 5% limit is 5.318. Our result is not significant; we have not been able to show any linear relation between the dowser's guess and the true position.

E. If we make 843 tests, each at the 5% level, we would expect about $\frac{5 \cdot 843}{100} = 42$ of the tests to be significant, even if all null hypotheses are true.

2.1 A Minitab printout is as follows:

Regression Analysis: Yield versus Fertil; Rain

```
The regression equation is
Yield = 28.1 + 0.0381 Fertil + 0.833 Rain

Predictor      Coef    SE Coef       T       P
Constant     28.095      2.491   11.28   0.000
Fertil     0.038095   0.005832    6.53   0.003
Rain         0.8333     0.1543    5.40   0.006

S = 2.31455    R-Sq = 98.1%    R-Sq(adj) = 97.2%

Analysis of Variance

Source           DF       SS        MS        F       P
Regression        2  1128.57    564.29   105.33   0.000
Residual Error    4    21.43      5.36
Total             6  1150.00

Source   DF   Seq SS
Fertil    1   972.32
Rain      1   156.25
```

The fit is good (R-square=98.1%). The yield would increase by 1 unit for 0.038 units increase in fertilizer, and by 1 unit for 0.8333 units increase in rain. Both fertilizer ($p = 0.003$) and rain ($p = 0.006$) are significantly related to yield.

A set of diagnostic plots reveals no major deviations from normality or homoscedasticity, but this is of course difficult to see in this small sample:

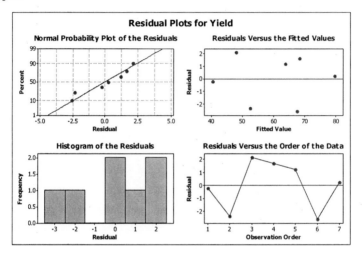

2.2 A.

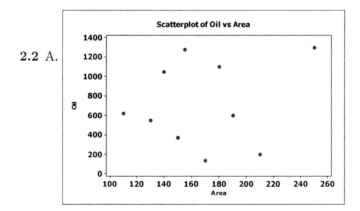

Regression Analysis: Oil versus Area

```
The regression equation is
Oil = 405 + 1.88 Area
```

B.
```
Predictor    Coef    SE Coef      T       P
Constant    404.9     633.6     0.64   0.541
Area        1.876     3.664     0.51   0.623

S = 451.720    R-Sq = 3.2%    R-Sq(adj) = 0.0%

Analysis of Variance
Source           DF       SS       MS      F      P
Regression        1    53485    53485   0.26  0.623
Residual Error    8  1632405   204051
Total             9  1685890
```

C. $r = \sqrt{0.032} = 0.178\,89$.

D. There is no significant relation between area and oil consumption, in this model. This seems strange!

Regression Analysis: Oil versus Temp

```
The regression equation is
Oil = 1002 - 51.7 Temp
```

E.
```
Predictor     Coef    SE Coef      T       P
Constant   1001.86      67.26  14.90   0.000
Temp        -51.724      7.350  -7.04  0.000

S = 171.191    R-Sq = 86.1%    R-Sq(adj) = 84.4%

Analysis of Variance

Source           DF       SS        MS      F       P
Regression        1  1451440   1451440  49.53  0.000
Residual Error    8   234450     29306
Total             9  1685890
```

There is a significant relation between oil consumption and temperature: when temperature increases on degree, oil consumption decreases by 51.724 units.

Regression Analysis: Oil versus Temp; Area

```
The regression equation is
Oil = 439 - 54.5 Temp + 3.43 Area

Predictor     Coef    SE Coef      T       P
Constant     438.7      129.5   3.39   0.012
Temp        -54.468     4.009 -13.59   0.000
Area         3.4304    0.7573   4.53   0.003

S = 92.2983    R-Sq = 96.5%    R-Sq(adj) = 95.5%
```

F.
```
Analysis of Variance

Source           DF       SS       MS      F       P
Regression        2  1626257  813129  95.45  0.000
Residual Error    7    59633    8519
Total             9  1685890

Source   DF   Seq SS
Temp      1  1451440
Area      1   174817
```

G. Yes, both area and temperature are significantly related to oil consumption.

H. Analysis 1: one m² more of area increases oil consumption by 1.876 units.

Analysis 2: one m² more of area increases oil consumption by 3.4304 units. This value is probably more correct since the model accounts also for temperature.

I. $\hat{y} = 438.7 - 54.468 \cdot (-1.5) + 3.4304 \cdot 200 = 1206.5$.

J. Possible other factors: Thickness of the glass; other insulation; amount of sunshine.

2.3 A. The best single predictor is Waist (highest R^2, highest adj R^2 of models with only one x variable.

B. The model with ht, waist is the best two-variable model (highest R^2, highest adj R^2).

C. Using adj R^2 as a criterion, the model with all three variables and the model with only ht, waist (the same adj R^2, namely 0.870). In such cases, choose the simplest of the models.

D. If you use the full model, the predicted value is 212.175.

If you only use ht and waist, the predicted value is 211.27

3.1 A. The hypothesis to test is apparently H_0: $\beta = 0.75$ against H_1: $\beta \neq 0.75$. This can be done as

$$t = \frac{b - 0.75}{s.e.(b)} = \frac{1.004 - 0.75}{0.0182} = 13.956.$$

There were 89 rats so t has $n - 2 = 87$ d.f.. Our value of t is highly significant, $p < 0.0001$.

B. Testing H_0: $\rho = 0$ is (for simple linear models) the same as testing H_0: $\beta = 0$. This test is

$$t = \frac{b - 0}{s.e.(b)} = \frac{1.004 - 0}{0.0182} = 55.156$$

on 87 d.f.. This is of course highly significant.

C. The predicted logged oxygen uptake is $\hat{y} = \log(1.715) + 1.004 \log(Age)$, i.e. $\hat{y} = 0.2343 + 1.004x$. For a rate aged 15 days, $x = 1.1761$. The predicted logged uptake is then $\hat{y} = 0.2343 + 1.004 \cdot 1.1761 = 1.4151$. A 95% prediction interval is

$$\hat{y} \pm t_{0.975,87} \sqrt{s_e^2 \left(1 + \frac{1}{n} + \frac{(x_0 - \bar{x})^2}{\sum (x_i - \bar{x})^2}\right)}$$

$$1.4152 \pm 1.98\sqrt{0.01\left(1 + \frac{1}{89} + \frac{(1.1761 - 0.83)}{56.2877}\right)}$$

or 1.4152 ± 0.20. The limits are 1.2152 and 1.6152. Taking antilogs, the corresponding limits for VO2 are 16.41 and 41.229. This means that 95% of the rats have an oxygen uptake between 16.41 and 41.23.

3.2 A. The model can be linearized using a log transformation. We obtain $\log(y) = \log(\alpha) + x \cdot \log(\beta)$ which is a linear regression with $\log(y)$ as dependent variable. A Minitab output (using natural log) is as

Regression Analysis: log(y) versus x

```
The regression equation is
log(y) = - 6.230 + 0.4530 x

S = 0.0490724    R-Sq = 99.9%    R-Sq(adj) = 99.9%

Analysis of Variance

Source        DF       SS          MS         F         P
Regression    1        14.3619     14.3619    5963.97   0.000
Error         4        0.0096      0.0024

Total         5        14.3715
```

follows:
model has a very good fit to the data: $R^2 = 99.9\%$.

B. We assume that the residuals in the log-transformed model are independent and approximately normally distributed.

C. For $x = 11$ the model gives $\log(y) = -6.230 + 0.4530 \cdot 11 = -1.247$. This gives the corresponding weight as $e^{-1.247} = 0.287$.

4.1 A. Statement a) means that there is interaction between variety and plant density; this is included in Model 4. Statement b) means that yield does not depend on variety; this is the case for model 1.

B. Model 3 assumes parallel lines while Model 4 does not. We compare these models: the difference in SS_e is 412.54157 on 5 d.f.. The hypothesis that the lines are parallel can be tested as $F = \frac{412.54157/5}{105.41604} = 0.78269$. This is clearly not significant, the hypothesis of parallel lines cannot be rejected.

C. The ANOVA table is:

Source	d.f.	SS	MS
Stand	1	12161.1673	12161.1673
Variety	5	3653.6524	730.7305
Block	3	1502.3944	500.7981
Error	14	1361.2859	97.2347
Total	23	18678.5	

D. The hypothesis that the average yield is the same is tested using $F = \frac{730.7305}{97.2347} = 7.5151$. Limits for F on (5,14) d.f. are 2.96 (5%); 4.695 (1%) and 7.922 (0.1%). The result is significant at the 1% level; the p value by computer is 0.0013.

4.2 A. We compare models 1 and 2: Change in $SS_e = 153 - 142 = 11$ on $5 - 3 = 2$ d.f.. We test the hypothesis of parallel lines as $F = \frac{11/2}{142/54} = 2.09$. The 5% limit is about 3.17. The result is not significant.

B. We compare models 2 and 3: Change in SSe$= 260 - 153 = 107$ on $3 - 1 = 2$ d.f.. We test the hypothesis of no differences between areas as $F = \frac{107/2}{153/56} = 19.582$. The 0.1% limit is about 7.83. The result is highly significant.

C. Since the hypothesis of parallel lines cannot be rejected, Model 2 provides a good summary of the data.

D. The regression lines for the three areas are:
Area 1: $y = 7.0 + 0.7x$
Area 2: $y = 4.6 + 0.7x$
Area 3: $y = 3.5 + 0.7x$
A plot of this relation is as follows:

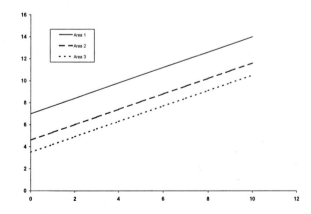

4.3 A. Average weekly growth is the coefficient for "Age", i.e. 0.48676.

B. Yes: the coefficient for the dummy variable D1 is -1.9184 which is significantly different from 0 ($p = 0.000$).

C. The average weight difference is equal to the difference between the coefficients of the dummy variables D1 and D2. Turkeys from Georgia weigh $-1.9184 - (-2.1919) = 0.2735$ more than turkeys from Virginia.

D. $\widehat{y} = 1.4309 + 0.48676 \cdot 24 - 2.1919 = 10.921$.

E.

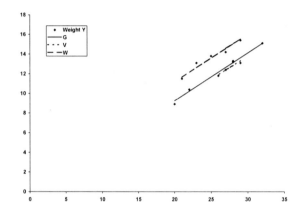

4.4 A. The samples are large so all hypotheses can be tested as $z = \frac{Parameter}{(s.e.)}$. Since physicians are the baseline group, we can test the hypothesis that firefighters and physicians have the same average air capacity as $z = \frac{-180}{54} = -3.33$. This is highly significant ($p = 0.0008$).

B. The air capacity for chemical workers and farm workers can be estimated as

  Chemical: $4500 - 39 \cdot \text{Age} - 9.0 \cdot \text{Smok} - 350$
  Farm: $4500 - 39 \cdot \text{Age} - 9.0 \cdot \text{Smok} - 380$
  Difference: 30

i.e. farm workers have 30 units lower air capacity than chemical workers, on the average.

C. $\widehat{y} = 4500 - 39 \cdot 50 - 9 \cdot 15 - 180 = 2235$.

4.5 There are some problems with these data, which makes the question rather realistic. Patient 1 occurs twice with the same data, and patient 2 also occurs twice but with different ages. I removed the duplicates (the second case of each) before analysis.

The model could either analyze the change ($y_1 - y_0$), or use y_1 as dependent variable and include y_0 as a covariate. The second approach is theoretically better.

Another question is how we should use the dose x. Patients randomized to B all got x, but some patients randomized to A also got some x if it was needed at the operation. This is known as "noncompliance": some patients were not treated as they were randomized. One can argue as follows: If some patients with treatment A still need some x, then this will be the case also for future patients with this treatment. Then, we actually compare the following treatments:

A: New treatment; add some x if needed.

B: Old treatment; give a standard dose of x.

With this way of reasoning, the dose x should not be included in the model.

A first model for y1 should also contain interactions. We can write it as

```
y1 = y0 treat age y0*treat age*treat y0*age;
```

The output from Proc GLM using this model is as follows:

```
Dependent Variable: y1

                                   Sum of
Source                    DF      Squares      Mean Square    F Value    Pr > F
Model                      6    51068.01760     8511.33627      12.21    <.0001
Error                     33    23001.95740      697.02901
Corrected Total           39    74069.97500

              R-Square     Coeff Var      Root MSE      y1 Mean
              0.689456     22.85333       26.40131      115.5250

Source                    DF    Type III SS    Mean Square    F Value    Pr > F
y0                         1     2512.546340    2512.546340      3.60    0.0664
Treat                      1      376.584455     376.584455      0.54    0.4675
age                        1     1149.563827    1149.563827      1.65    0.2080
y0*Treat                   1      668.185130     668.185130      0.96    0.3347
age*Treat                  1      660.763708     660.763708      0.95    0.3373
y0*age                     1     1255.099440    1255.099440      1.80    0.1888
```

There seem to be no significant effects of the treatment. However, a normal plot of the residuals indicates non-normality:

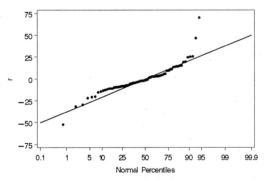

A similar model was therefore tested on log-transformed data. (y_0 was also log-transformed). This model showed a better fit to the data

($R^2 = 0.868$, compared with $R^2 = 0.689$ for the first model. The normal plot was slightly better, but still not good. The results from the analysis of logged y values was as follows:

```
Dependent Variable: ly1

                                  Sum of
Source                    DF     Squares      Mean Square    F Value    Pr > F
Model                      6   6.65342622      1.10890437      77.23    <.0001
Error                     70   1.00514562      0.01435922
Corrected Total           76   7.65857184

            R-Square     Coeff Var      Root MSE      ly1 Mean
            0.868755     2.519314       0.119830      4.756452

Source                    DF   Type III SS    Mean Square    F Value    Pr > F
ly0                        1   0.03963171      0.03963171       2.76    0.1011
Treat                      1   0.00012313      0.00012313       0.01    0.9265
age                        1   0.00828548      0.00828548       0.58    0.4500
ly0*Treat                  1   0.00009567      0.00009567       0.01    0.9352
age*Treat                  1   0.00130843      0.00130843       0.09    0.7637
ly0*age                    1   0.00786421      0.00786421       0.55    0.4617
```

As a conclusion, we have not been able to demonstrate any effect of the treatment.

5.1 A. H_0: $p = 0.5$ H_1: $p > 0.5$. Out of $n = 7$ persons, all ($x = 7$) gave a positive difference. The probability that this should happen if H_0 is true may be calculated using the binomial distribution with $n = 7$, $p = 0.5$ and $x = 7$. The p value is 0.0078. This is significant.
B. A matched t test gives $t = \frac{\bar{d}}{\sqrt{\frac{s_d^2}{n}}} = \frac{0.6771}{\sqrt{\frac{0.561}{7}}} = 2.3918$. The 5% limit for t on 6 d.f. is 2.447 (double-sided) or 1.943 (single-sided). The test is significant if we have chosen a single-sided test, but not if we choose a double-sided test

5.2 A nonparametric alternative to ANOVA with blocks is Friedman's test. When each block contains only two observations, this is equivalent to a sign test. Thus, we look at the signs of the differences between the two types of plants:

Pair	Cross-fertilized	Self-fertilized	Difference	Sign
1	23.5	17.4	6.1	+
2	12.0	20.4	-8.4	-
3	21.0	20.0	1.0	+
4	22.0	20.0	2.0	+
5	19.1	18.4	0.7	+
6	21.5	18.6	2.9	+
7	22.1	18.6	3.5	+
8	20.4	15.3	5.1	+
9	18.3	16.5	1.8	+
10	21.6	18.0	3.6	+
11	23.3	16.3	7.0	+
12	21.0	18.0	3.0	+
13	22.1	12.8	9.3	+
14	23.0	15.5	7.5	+
15	12.0	18.0	-6.0	-

We want to test if the median difference is zero, i.e. H_0: $Md = 0$, against the single-sided alternative H_1: $Md > 0$. Out of the 15 differences, 13 are positive. If H_0 is true, the probability of a positive difference is 0.5. The prob-value of the test is computed using a binomial distribution with $n = 15$ and $p = 0.5$. It is

$$\begin{aligned}\text{prob value} &= P(x=13) + P(x=14) + P(x=15) \\ &= 0.0032043 + 0.0004578 + 0.0000305 \\ &= 0.0036926\end{aligned}$$

The null hypothesis can be rejected since the prob value is smaller than 5% (and even smaller than 1%).

The assumption underlying the analysis is that the 15 pairs of plants are independent of each other.

5.3 H_0: the medians are equal for the two drugs. H_0: the medians are not equal

The test can be done using a Wilcoxon-Mann-Whitney test. The ranks of the observations are already given in the data table. We calculate

$$\begin{aligned}\overline{R}_1 &= 25/6 = 4.167 \\ \overline{R}_2 &= 66/7 = 9.429.\end{aligned}$$

Test statistic:
$$\chi^2 = \frac{12 \cdot 6 \cdot 7 \cdot (4.167 - 9.429)^2}{13^2 \cdot 14} = 5.90$$

This should be compared with the 5% limit for χ^2 on 1 d.f. which is 3.84. Our result is significant. (The computer calculated p value is $p = 0.015$). It seems that the drugs have different effects on median reaction time.

6.1 H_0: $p = 0.75$, H_1: $p \neq 0.75$. The hypothesis can be tested using a χ^2 test.

If H_0 is true, we should on the average get $0.75 \cdot 100 = 75$ fish with translucent scales and 25 with pigmented scales. We get the following table of observed and expected frequencies:

	Translucent	Pigmented	Total
O_i	90	10	100
E_i	75	25	100

We calculate $\chi^2 = \sum \frac{(O_i - E_i)^2}{E_i} = \frac{(90-75)^2}{75} + \frac{(10-25)^2}{25} = 3 + 9 = 12$. This should be compared with the 5% limit for χ^2 on $2 - 1 = 1$ d.f. The limit is 3.84. Our observed value is larger than this limit; it is also larger than the 1% limit (6.635) and the 0.1% limit (10.828). The p value is smaller than 0.001. This is highly significant and we reject the hypothesis that $p = 0.75$.

6.2 H_0: $p = 15/16 = 0.9375$, H_1: $p \neq 0.0.9375$. This can be tested using a χ^2 test. If H_0 is true we should expect $0.9375 \cdot 100 = 93.75$ fish with translucent scales and 6.25 with pigmented scales. We get the following table of observed and expected numbers:

	Translucent	Pigmented	Total
O_i	90	10	100
E_i	93.75	6.25	100

We calculate $\chi^2 = \sum \frac{(O_i - E_i)^2}{E_i} = \frac{(90-93.75)^2}{93.75} + \frac{(10-6.25)^2}{6.25} = 0.15 + 2.25 = 2.40$. This is compared with the 5% limit for χ^2 on $2 - 1 = 1$ d.f.. The limit is 3.84. Our value is smaller than this limit, and we cannot reject H_0. Our data agree better with the hypothesis in this exercise than in the previous one. In both questions, the sample is large enough for approximation with a χ^2 distribution (all E_i are at least 5). We can hope that the samples are independent, but this cannot be checked.

6.3 H_0: The experimental group and the control group have the same lifetime distribution. The hypothesis can be tested using a χ^2 test. If H_0

is true, the expected numbers can be calculated as (row sum)(column sum)/n. We get the following table of observed and expected number; expected numbers are given in parenthesis.

	Exp. group	Control group	Total
At least 3 years	108 (92.16)	36 (51.84)	144
less than 3 years	36 (51.84)	45 (29.16)	81

We calculate $\chi^2 = \sum \frac{(O_{ij}-E_{ij})^2}{E_{ij}} = \frac{(108-92.16)^2}{92.16} + \frac{(36-51.84)^2}{51.84} + \frac{(36-51.84)^2}{51.84} + \frac{(45-29.16)^2}{29.16} = 2.7225 + 4.8400 + 4.8400 + 8.6044 = 21.01$. The 5% limit for χ^2 on $(2-1)(2-1) = 1$ d.f. is 3.84. The observed value is larger than this limit; it is also larger than the 1 limit 6.635 and the 0.1% limit 10.828. The p value is smaller than 0.001. This is highly significant: there seems to be a relation between survival and the nutrient.

6.4 H_0: the treatments do not affect the degree of coloring of the potatoes. The hypothesis can be tested using a χ^2 test. The expected numbers, assuming that H_0 is true, are (row sum)(column sum)/n We get χ^2 as follows, using Minitab:

Chi-Square Test

```
Expected counts are printed below observed counts

          a1b1      a1b2      a2b1      a2b2     Total
    1       56        64        36        38       194
          48.50     48.50     48.50     48.50

    2       45        36        44        48       173
          43.25     43.25     43.25     43.25

    3       18        13        27        20        78
          19.50     19.50     19.50     19.50

    4        6        12        18        19        55
          13.75     13.75     13.75     13.75

Total      125       125       125       125       500

Chi-Sq =   1.160 +   4.954 +   3.222 +   2.273 +
           0.071 +   1.215 +   0.013 +   0.522 +
           0.115 +   2.167 +   2.885 +   0.013 +
           4.368 +   0.223 +   1.314 +   2.005 =   26.518
DF = 9,  P-Value = 0.002
```

The result is strongly significant; the different treatments have different effect on the color of potatoes.

6.5 We will test the hypothesis that lung cancer is independent of owning a pet bird. If independence holds, the expected numbers can be calculated as $E_{ij} = \frac{n_{i.} n_{.j}}{n}$. We get: $E_{11} = \frac{199 \cdot 234}{668} = 69.71$, $E_{12} =$

$\frac{199 \cdot 429}{668} = 127.80$, $E_{21} = \frac{469 \cdot 234}{668} = 164.29$ and $E_{22} = \frac{469 \cdot 429}{668} = 301.20$. χ^2 can be calculated as $\chi^2 = \sum \frac{(O_{ij} - E_{ij})^2}{E_{ij}} = \frac{(98 - 69.71)^2}{69.71} + \frac{(101 - 127.80)^2}{127.80} + \frac{(141 - 164.29)^2}{164.29} + \frac{(328 - 301.20)^2}{301.20} = 22.79$. Limits for χ^2 on $(2-1)(2-1) = 1$ d.f. are 3.84 (5%), 6.64 (1%) and 10.83 (0.1%). Our value of chi-square is larger than all these limits and thus "three star" significant. There is some relation between lung cancer and pet birds. Assumptions: Independent samples from the two populations of patients; the expected numbers should be large (>10). The latter assumption is fulfilled, but we don't have any information on the sampling procedure.

6.6 A. H_0: The blood group distribution is the same as in the US. We calculate expected numbers as $E_i = n \cdot p_i$, where p_i are the proportions in the US population:
$E_O = 500 \cdot 0.45 = 225$; $E_A = 500 \cdot 0.40 = 200$; $E_B = 500 \cdot 0.11 = 55$ and $E_{AB} = 500 \cdot 0.04 = 20$. The hypothesis is tested using

$$\chi^2 = \sum \frac{(O_i - E_i)^2}{E_i}$$
$$= \frac{(228 - 225)^2}{225} + \frac{(206 - 200)^2}{200} + \frac{(40 - 55)^2}{55} + \frac{(26 - 20)^2}{20} = 6.11.$$

This is compared with the limit for χ^2 on $(k-1) = 3$ degrees of freedom. The 5% limit is 7.815; we have not been able to show any differences in blood group distribution between this ethnic group and the US population.

B. H_0: The blood group distribution is the same for Rh+ as for Rh−. We have observed numbers

	O	A	B	AB	Sum
Rh +	198	176	28	22	424
Rh −	30	30	12	4	76
Sum	228	206	40	26	500

Expected numbers are calculated as $E_{ij} = \frac{(\text{Sum in row } i)(\text{Sum in column } j)}{n}$. They are:

	O	A	B	AB	Sum
Rh +	193.34	174.69	33.92	22.05	424
Rh −	34.66	31.31	6.08	3.95	76
Sum	228	206	40	26	500

To test H_0 we calculate $\chi^2 = \sum \frac{(O_{ij} - E_{ij})^2}{E_{ij}} = 7.60$. This should be compared with χ^2 on $(r-1)(c-1) = (4-1)(2-1) = 3$ d.f.. The 5% limit is 7.815; our result is not quite significant at the 5% level. We have not been able to demonstrate any differences in blood-group distribution between RH+ and Rh− for this ethnic group.

6.7 A. H_0: The blood-group distribution is the same in the two groups. We calculate expected numbers as $E_{ij} = \frac{(\text{Sum in row } i)(\text{Sum in column } j)}{n}$. They are:

Blood type	Ulcer patients	Controls
O	779.43	4709.57
A	681.31	4116.69
B	143.99	870.01
AB	50.27	303.73
Total	1655.00	10000.00

The hypothesis is tested using

$$\chi^2 = \sum \frac{(O_i - E_i)^2}{E_i}$$

$$= \frac{(911-779.43)^2}{779.43} + \frac{(579-681.31)^2}{681.31} + \frac{(124-143.99)^2}{143.99}$$
$$+ \frac{(41-50.27)^2}{50.27} + \frac{(4578-4709.57)^2}{4709.57} + \frac{(4219-4116.69)^2}{4116.69}$$
$$+ \frac{(890-870.01)^2}{870.01} + \frac{(313-303.73)^2}{303.73}$$
$$= 49.2$$

This is compared with the limit for χ^2 on $(r-1)(c-1) = 3$ degrees of freedom. The 5% limit is 7.815 and the 0.1% limit is 16.266. The result is significant; the two groups seem to differ in blood group distribution.

7.1 A. To test whether smoking has any effect we calculate $z = \frac{Estimate}{s.e.} = \frac{0.5640}{0.2871} = 1.9645$ which is just above the 5% level for z, which is 1.96. Thus, the result is significant at the 5% level.

B. For smokers we get $\log\left(\frac{p}{1-p}\right) = -4.0686 + 0.5640 = -3.5046$. This gives $\frac{p}{1-p} = e^{-3.5046} = 0.030059$ so $p = 0.030059 - 0.030059p$; $1.030059p = 0.030059$; $p = \frac{0.030059}{1.030059} = 0.029182$.

For non-smokers, $\log\left(\frac{p}{1-p}\right) = -4.0686$. This gives $\frac{p}{1-p} = e^{-4.0686} = 0.017101$ so $p = 0.017101 - 0.017101p$; $1.017101p = 0.017101$; $p = \frac{0.017101}{1.017101} = 0.016813$. The risk for smokers is nearly twice as large as the risk for non-smokers, although both risks are small.

7.2 A. H_0: temperature has no effect is tested as $z \approx \frac{Estimate}{(s.e.)} = \frac{-0.1034}{0.0430} = -2.4047$. The 5% limit is ± 1.96 so our result is significant at least at the 5% level. The p value is 0.016. Note, however, that the sample is rather small so the Normal approximation is a bit doubtful.

B. The model is $\log(\mu) = \beta_0 + \beta_1 x$ i.e. $\log(\mu) = 5.9691 - 0.1034 \cdot 31 = 2.7637$. This gives $\mu = e^{2.7637} = 15.86$

C. We use the Poisson distribution. It is easiest to calculate $P(0)$, $P(1)$ and $P(2)$ and add them together:

$P(0) = \frac{\mu^x e^{-\mu}}{x!} = \frac{15.86^0 e^{-15.86}}{0!} = 0.0000001$.

$P(1) = \frac{15.86^1 e^{-15.86}}{1!} = 0.0000021$.

$P(2) = \frac{15.86^2 e^{-15.86}}{2!} = 0.0000163$.

These add up to 0.0000185, so the probability that three or more fail is $1 - 0.0000163 = 0.9999837$. If the model holds for this temperature, it is nearly certain that three or more O-rings will break!

7.3 The data may be analyzed as a log-linear model. Analysis using Proc GGENMOD in SAS reveals that the three-way interaction is not significant. A model without that interaction gave the following result:

LR Statistics For Type 3 Analysis

Source	DF	Chi-Square	Pr > ChiSq
A	1	14.76	0.0001
B	1	0.10	0.7562
y	1	110.20	<.0001
A*B	1	0.00	0.9466
y*A	1	18.75	<.0001
y*B	1	0.12	0.7269

It appears that the degree of miscoloring (y) is significantly related to nitrogen (factor A) but not to phosphorus (factor B). Two cross. tables may illustrate the result:

Table of A by y

Frequency Row Pct	1	2	3	4	Total
1	120 / 48.00	81 / 32.40	31 / 12.40	18 / 7.20	250
2	74 / 29.60	92 / 36.80	47 / 18.80	37 / 14.80	250
Total	194	173	78	55	500

Table of B by y

Frequency Row Pct		1	2	3	4	Total
	1	92 36.80	89 35.60	45 18.00	24 9.60	250
	2	102 40.80	84 33.60	33 13.20	31 12.40	250
Total		194	173	78	55	500

Miscoloring of grade 1 is more common when A=1 than when A=2.

B. The data were fitted using a generalized linear model with a multinomial distribution and a cumulative logit link. Part of the results are as follows:

LR Statistics For Type 3 Analysis

Source	DF	Chi-Square	Pr > ChiSq
A	1	21.71	<.0001
B	1	0.58	0.4481
A*B	1	0.01	0.9411

Level of nitrogen (factor A) is significantly related to miscoloring. Neither factor B (phosphorus), nor the A*B interaction, have any discernible effect. To interpret the direction of the relation we need to know whether low or high values of y denote large miscoloring.

7.4 A. Both models use a binomial distribution and the logit link $g(\mu) = \log \frac{p}{1-p}$. Notation: species=$x$, time=$t$, temp=$z$, humidity=$u$. We use p =probability of dying. The models may be written

$$\text{logit}(p) = \log\left(\frac{p}{1-p}\right) = \mu + \beta_1 x + \beta_2 t + \beta_3 z + \beta_4 u \text{ (model 0) and}$$

$$\text{logit}(p) = \log\left(\frac{p}{1-p}\right) = \mu + \beta_1 x + \beta_2 t + \beta_3 z + \beta_4 u + \beta_5 \cdot x \cdot t$$
$$+ \beta_6 \cdot x \cdot z + \beta_7 \cdot x \cdot u.$$

B. i) Model 0 is nested within Model 1 so the test can be made by comparing the model deviances. It holds that $D_0 - D_1$ is asymptotically Chi-2-with d.f. = the difference between the models d.f. Here,

$$D_0 - D_1 = 55.07 - 53.99 = 1.08$$

with $91 - 88 = 3$ d.f. This is not significant (5% limit 7.815). The three extra parameters in model 1 do not improve the model fit. We may select model 0.

ii) For model choice we often use the AIC. For model 0, $AIC = 223.93$, while Model 1 gives $AIC = 228.85$. Model 0 has the smallest AIC and should be selected.

Note that the value of Deviance/d.f. cannot be used to compare competing models.

C. According to the printout we estimate the parameter β_2 for the time effect with $\hat{\beta}_2 = 1.503$ with estimated standard error=0.102. A confidence interval for β_2 is obtained as

$$1.503 \pm z \cdot 0.102. \text{ With } z = 1.96 \text{ the 95\% confidence interval is}$$
$$1.503 \pm 1.96 \cdot 0.102;\ 1.503 \pm 0.20;\ 1.303\text{—}1.703.$$

D. Yes. The rest gives $z = 8.005$ which is significant ($p < 0.001$).

E. In our models, p = probability that a snail dies. Since the estimated coefficient $\hat{\beta}_1$ for species is positive, the risk of dying is larger for species 1 than for species 0. Thus, species 0 has higher survival than species 1.

F. The odds ratio can be estimated as $e^{\hat{\beta}_1} = e^{1.309} = 3.7025$ when species 1 is compared with species 0.

G. If the variables had been "class" variables, their degrees of freedom should have been (number of classes-1). However, they all have only 1 degree of freedom. This shows that they were included as numeric variables.

8.1 A. The two-way ANOVA does not account for the fact that observations within individual are dependent. Therefore, it is not a valid analysis.

B. The p values will normally be smaller using the incorrect analysis. Thus, it will be too easy to get a significant result.

C. A better analysis would be to use a mixed model where "time" is included as a class variable, and where the covariance structure is left free (unstructured). The model should contain method, time, method*time and use time as repeated measures variable.

8.2 A. The analysis is supposed to be a "repeated-measures ANOVA", i.e. a so called "split-plot" analysis. However, it is not. If you calculate the mean squares you will notice that all tests were made using MS_e in the denominator. This is not correct for this type of analysis: the treatment factor should be tested as explained in B.

Source	SS	d.f.	MS	F test against Error	Correct test
Treat	1.020	1	1.0200	17.65	0.983
Trees	5.187	5	1.0374	17.95	
Date	7.646	4	1.9115	33.07	
Date*Treat	0.349	4	0.0873	1.51	
Error	1.156	20	0.0578		
Total		34			

B. The correct test of the "main plot factor" (i.e. of the treatment effect) is to use the trees within treatment mean square in the denominator. This gives the F ratio

$$F_{1,5} = \frac{MS_{treat}}{MS_{trees(treat)}} = \frac{1.02}{1.0374} = 0.983.$$

This is not significant at any reasonable level. Thus, contrary to the published result, there is no significant treatment effect. Use the test in the column "Correct test" in the ANOVA table, not the test against error.

C. We have not recommended the "spli- plot" approach for analysis of repeated measures data. It assumes that the correlations between time points is constant, regardless of the time distance. A better alternative is to use a mixed model. The covariance structure may be chosen as unstructured or, if there is need for simplification, e.g. as an AR(1) structure.

One might also question the normality assumption. The raw data seem to be counts of insects. For such data, a Poisson distribution may be preferred. Thus, a generalized linear model with a Poisson distribution might be used, either as a mixed model or using the GEE approach to take care of the time dependence.

8.3 We start by graphing the data. A plot of the mean values over time is given in Figure A.7. It seems that group 1 had a higher value already at $t = 0$ and that the three other experimental groups start at about the same level but that they diverge over time. A linear trend does not seem useful. Several analyses are possible to investigate this.

Note that the sample sizes were incorrectly given in the question. The data reveal that the experiment is unbalanced, contrary to the statement in the text.

Analysis using mixed models

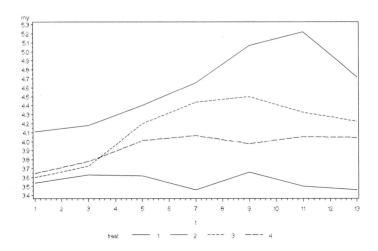

Figure A.7: Mean profile plot for the example data.

My preferred analysis of these data is to use a mixed model. For the baseline model I used unstructured covariances. This model has a slightly better fit (AIC=292.7) than the AR(1) model (AIC=296.3). Program:
```
PROC MIXED data=dogs;
CLASS dog treat t;
MODEL y = t treat t*treat;
REPEATED / subject = dog*treat type=UN;
RUN;
```

Output:

```
                  Fit Statistics
       -2 Res Log Likelihood        236.7
       AIC (smaller is better)      292.7
       AICC (smaller is better)     301.0
       BIC (smaller is better)      337.0
           Type 3 Tests of Fixed Effects
                   Num    Den
       Effect      DF     DF    F Value    Pr > F
       t            6     32      8.00     <.0001
       treat        3     32      6.91     0.0010
       treat*t     18     32      2.37     0.0162
```

To simplify the model we might want to subdivide the time trend into polynomial terms. However, the linear, quadratic and cubic (etc.) models all have a worse fit (in terms of AIC) than the model with time

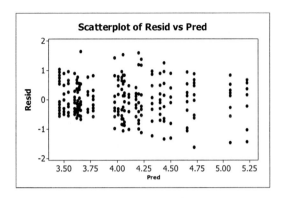

Figure A.8: Plot of residuals versus fitted values.

as a class variable, so we stick to the baseline model (see Table A.6).

Table A.6: AIC for models where time is modeled by a polynomial.

Degree	AIC
1	306.4
2	337.6
3	363.2
4	414.7
5	478.5
6	545.4

As we suspected from the graph, there is a significant effect of time, a significant treatment effect and a significant treatment*time interaction (since the curves are not parallel). In a detailed analysis, we might go on and compare individual time points etc., but this is not done here.

As a check of the assumptions, we should plot the residuals against $\hat{y}$ (to check if the variance is constant), and do a normal probability plot (to check normality). Such plots, made using Minitab, are given in Figures A.8 and A.9. Figure A.8 does not indicate any problems with heteroscedasticity, and figure A.9 shows an unusually good agreement with a normal distribution.

Repeated-measures ANOVA

An alternative approach is to use the "repeated-measures ANOVA" feature that is available in many statistics packages. This analysis, using the GLM procedure of SAS, gives the following results.

The assumption that the correlations between occasions is constant

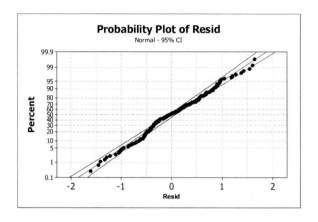

Figure A.9: Normal probability plot of the residuals.

("spherical") does not seem to be tenable:

```
                        Sphericity Tests

                              Mauchly's
   Variables            DF    Criterion    Chi-Square    Pr > ChiSq

   Transformed Variates 20    0.0320293    102.46839     <.0001
   Orthogonal Components 20   0.0320293    102.46839     <.0001
```

The effects of time, time*treatment and treatment are all significant. However, because of the non-sphericity, the p values are somewhat doubtful.

```
   Manova Test Criteria and Exact F Statistics for the Hypothesis of no time Effect
                     H = Type III SSCP Matrix for time
                          E = Error SSCP Matrix

                        S=1     M=2     N=12.5

   Statistic                    Value    F Value   Num DF   Den DF   Pr > F

   Wilks' Lambda              0.40000186    6.75      6        27    0.0002
   Pillai's Trace             0.59999814    6.75      6        27    0.0002
   Hotelling-Lawley Trace     1.49998835    6.75      6        27    0.0002
   Roy's Greatest Root        1.49998835    6.75      6        27    0.0002
                            The GLM Procedure
                    Repeated Measures Analysis of Variance

   Manova Test Criteria and F Approximations for the Hypothesis of no time*Treat
Effect
                     H = Type III SSCP Matrix for time*Treat
                          E = Error SSCP Matrix

                        S=3     M=1     N=12.5

   Statistic                    Value    F Value   Num DF   Den DF   Pr > F

   Wilks' Lambda              0.35243223    1.90     18      76.853   0.0277
   Pillai's Trace             0.83745460    1.87     18         87    0.0290
   Hotelling-Lawley Trace     1.33228759    1.93     18      48.296   0.0357
   Roy's Greatest Root        0.78840094    3.81      6         29    0.0064
```

The GLM Procedure
Repeated Measures Analysis of Variance
Tests of Hypotheses for Between Subjects Effects

Source	DF	Type III SS	Mean Square	F Value	Pr > F
Treat	3	39.19686111	13.06562037	6.91	0.0010
Error	32	60.52964683	1.89155146		

A third-degree polynomial is needed to account for the trend. (Only the first three terms are listed, the others are not significant).

Contrast Variable: time_1

Source	DF	Type III SS	Mean Square	F Value	Pr > F
Mean	1	6.79760547	6.79760547	19.32	0.0001
Treat	3	4.82916270	1.60972090	4.58	0.0089
Error	32	11.25798016	0.35181188		

Contrast Variable: time_2

Source	DF	Type III SS	Mean Square	F Value	Pr > F
Mean	1	2.68832299	2.68832299	10.53	0.0028
Treat	3	0.92926058	0.30975353	1.21	0.3208
Error	32	8.16978704	0.25530584		

Contrast Variable: time_3

Source	DF	Type III SS	Mean Square	F Value	Pr > F
Mean	1	0.54512641	0.54512641	4.30	0.0463
Treat	3	1.43211111	0.47737037	3.76	0.0202
Error	32	4.05770370	0.12680324		

Other approaches

There are some alternatives to the above analyses. One approach would be to summarize the data for each dog into a few statistics and then analyze these using the "summary-measures" approach. The mean value or the sum of the observations for each dog, or the slope in a linear regression, are possible summary statistics. However, the graph indicates nonlinearity, so the linear regression approach is not recommended for these data.

9.1 A. Top red and Plendour are closest; they form one cluster.

B. and C. If the varieties are numbered 1, 2, 3, 4,, 5, the following analysis is obtained:

Step	Number of clusters	Similarity level	Distance level	Clusters joined		New cluster	Number of obs. in new cluster
1	4	76.8827	13.2	1	3	1	2
2	3	68.6515	17.9	4	5	4	2
3	2	59.8949	22.9	1	2	1	3
4	1	48.1611	29.6	1	4	1	5

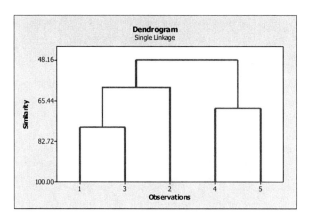

10.1 A. The reliability is calculated:: $k_0 = 2$, $s_t^2 = (BMS - WMS)/2 = 67.795$; $+ s_e^2 = 1.24$;
$R = (s_t^2)/(s_t^2 + s_e^2) = 67.795/(67.795 + 1.24) = 0.982$

B. $n = 2(s^2)(z_{0.975} + z_{0.90})^2/10^2 = 14.03$. We should use 15 patients/group..

C. $n = 2p(1-p)*(z_{0.975}+z_{0.80})^2/(0.1)^2 = 2\cdot 0.5\cdot 0.5\cdot (1.96+0.842)^2/0.01 = 392.56$. We need nearly 400 patients.

10.2 A. We summarize the data in an ANOVA table. The grand mean is

$$\overline{y}_{..} = \frac{10.4 + 14.5 + 12.2 + 11.8}{4} = 12.225.$$

$$SS_e = \sum (n_i - 1)\, s_i^2 = 4\,(5.2 + 6.7 + 5.7 + 6.1) = 94.8.$$

$$SS_A = \sum n_i\, (\overline{y}_{i.} - \overline{y}_{..})^2 = 5((10.4 - 12.225)^2 + (14.5 - 12.225)^2 +$$
$$(12.2 - 12.225)^2 + (11.8 - 12.225)^2) = 43.4375.$$

The ANOVA table is:

Source	SS	d.f.	MS
A: Persons	43.4375	3	14.4792
e: Residual	94.8000	16	5.9250
Total	138.2375	19	

The reliability is calculated as

$$R = \frac{MS_A - MS_e}{MS_A + (n-1)\,MS_e} = \frac{14.4792 - 5.9250}{14.4792 + (5-1)\,5.9250} = 0.22405.$$

(A rather low reliability!) Alternative calculation: $\widehat{\sigma}_e^2 = MS_e = 5.9250$,

$\hat{\sigma}_T^2 = \frac{MS_A - MS_e}{n} = \frac{14.4792 - 5.9250}{5} = 1.7108$, $R = \frac{\hat{\sigma}_T^2}{\hat{\sigma}_T^2 + \hat{\sigma}_e^2} = \frac{1.7108}{1.7108 + 5.9250} = 0.22405$.

B. We calculate the sample size based on $\alpha = 0.05$ (double-sided), $1 - \beta = 0.90$, $\Delta = 1$. We use the estimated variance from question A. The variance for one observation is $Var(y) = \sigma_T^2 + \sigma_e^2$ which is estimated as $1.7108 + 5.9250 = 7.6358$. We get

$$n = \frac{2\left(\sigma_T^2 + \sigma_e^2\right)\left(z_{1-\alpha/2} + z_{1-\beta}\right)^2}{\Delta^2}$$

$$= \frac{2 \cdot 7.6358(1.96 + 1.282)^2}{1^2} = 160.5$$

We should choose $n \geq 161$.

10.3 We should use some method to allocate α to the interim analyses, for example O'Brien-Fleming. This gives p values 0.0006, 0.0151 and 0.0471 for the three analyses.
A. The test gives $z = -1.33$ so the p value is clearly above the first limit 0.0006. The trial should continue.
B. The test gives $z = -1.95$ which would have been "nearly" significant. But now we demand the p value 0.0151, which is not reached. The trial should continue.
C. The test gives $z = -2.14$; the double sided p value is $2 \cdot 0.016 = 0.032$ which is smaller than the required value 0.0471. The result is significant and suggests a difference between the treatments.

Note that e.g. Pocock's method would not have given a significant result.

10.4 The tests at each time point are regular two-sample t tests. Since the samples are rather large, the test statistics may be regarded as standard normal. We can choose to assume equal variances, but the tests work rather well even without that assumption. The test statistics are calculated:
A. For the first analysis we get $s^2 = 81$ and $t = \frac{104 - 100}{\sqrt{81\left(\frac{1}{25} + \frac{1}{25}\right)}} = 1.57$.
B. Analysis 2 gives $s^2 = 72.5$ and $t = \frac{103 - 99}{\sqrt{72.5\left(\frac{1}{50} + \frac{1}{50}\right)}} = 2.35$.
C. Analysis 3 (=the final one) gives $s^2 = 110.5$ and $t = \frac{102 - 98}{\sqrt{110.5\left(\frac{1}{75} + \frac{1}{75}\right)}} = 2.33$.

D. We should use a method that leaves a large part of α to the final analysis. Either of O'Brien-Fleming, Haybittle-Peto or Pocock can be used; I prefer the first one. With three analyses, O'Brien-Fleming gives the limits A: 3.438 (we got 1.57; not significant); B: 2.431 (we got 2.35;

not significant) and C: 1.985 (we got 2.33; significant). The trial should not be stopped under way, but in the end we get a significant difference between treatments.

10.5 For each occasion, we calculate the information fraction as (cumulative number of cases)/3837. These are used in the formula for Pockock's method:
$$\alpha_P(\tau) = \alpha \log\left[1 + \tau(e-1)\right].$$
where $e = 2.7183$. We get:

Cases	Cum. cases	Information	Pocock p	Z
131	131	0.0341	0.0029	2.98
59	190	0.0495	0.0041	2.87
688	878	0.2288	0.0166	2.40
796	1674	0.4353	0.0280	2.20
853	2527	0.6586	0.0378	2.08
904	3431	0.8942	0.0465	1.99
Total	3837			

The values of Z where the test is significant are given in the rightmost column. We see that the test statistic passes the limit at analysis 4, where $Z = 2.30$ while the limit is 2.20.

10.6 A. To test the hypothesis of equivalence we test two hypotheses:
H_{0A}: $p_1 - p_2 \geq 0.03$ against H_{1A}: $p_1 - p_2 < 0.03$.
H_{0B}: $p_1 - p_2 \leq -0.03$ against H_{1B}: $p_1 - p_2 > -0.03$.

For the joint test to have level 0.10, each of these two tests should be made at level 0.05. Since the tests are single-sided, the limit is $z = 1.645$ (with different signs in the two tests). The two tests are:

Test A. $z = \dfrac{\frac{61}{692} - \frac{57}{682} - 0.03}{\sqrt{\frac{61}{692}\left(1-\frac{61}{692}\right)/692 + \frac{57}{682}\left(1-\frac{57}{682}\right)/682}} = -1.6823$ which is significant; H_{0A} can be rejected.

Test B. $z = \dfrac{\frac{61}{692} - \frac{57}{682} + 0.03}{\sqrt{\frac{61}{692}\left(1-\frac{61}{692}\right)/692 + \frac{57}{682}\left(1-\frac{57}{682}\right)/682}} = 2.2873$ which is significant; H_{0B} can be rejected.

Since both H_{0A} and H_{0B} can be rejected, we have shown equivalence, at the level 0.10.

B. The sample size that is needed can be calculated from
$$n = \frac{\left[p_1(1-p_1) + p_2(1-p_2)\right](z_{1-\alpha} + z_{1-\beta})^2}{\left[\delta - (p_1-p_2)\right]^2}.$$

We get

$$n = \frac{[0.1(1-0.1) + 0.1(1-0.1)](1.645 + 1.282)^2}{[0.03]^2} = 1713.5$$

We need large samples: 1714 patients per group.

10.8 A. The distribution of "number of patients who get DLT" (out of $n = 3$) is binomial with $n = 3$, $p = 0.25$. Denote the number of patients with DLT with X. It holds that:
$P(X = 0) = \binom{3}{0} 0.25^0 \cdot 0.75^3 = 0.42188$.
$P(X = 1) = \binom{3}{1} 0.25^1 \cdot 0.75^2 = 0.42188$.
$P(X > 1) = 1 - 0.42188 - 0.42188 = 0.15624$.

If $X = 1$ we take three more patients. If one or more get DLT, dose escalation is stopped. The probability that this will happen is $(1 - 0.42188) = 0.57812$. The probability to first get $X = 1$, and then $X \geq 1$ is $0.42188 \cdot 0.57812 = 0.24390$. The total probability of DLT is thus $0.15624 + 0.24390 = 0.40014$.

B. In the same way we can write
$P(X = 0) = \binom{3}{0} p^0 \cdot (1-p)^3 = (1-p)^3$.
$P(X = 1) = \binom{3}{1} p^1 \cdot (1-p)^2 = 3p(1-p)^2$.
$P(X > 1) = 1 - (1-p)^3 - \left(3p(1-p)^2\right) = 3p^2 - 2p^3$.
The probability to get first $X = 1$, and then $X \geq 1$ is

$$\left(3p(1-p)^2\right)\left(1 - (1-p)^3\right) = 9p^2 - 27p^3 + 30p^4 - 15p^5 + 3p^6.$$

The total probability that dose escalation will be stopped is thus

$$\begin{aligned}&\left(3p^2 - 2p^3\right) + \left(9p^2 - 27p^3 + 30p^4 - 15p^5 + 3p^6\right) \\ &= 12p^2 - 29p^3 + 30p^4 - 15p^5 + 3p^6.\end{aligned}$$

C. The points are (0.25, 0.40); (0.50, 0.83) and (0.75, 0.98). This, together with (0, 0) and (1, 1) gives the following graph:

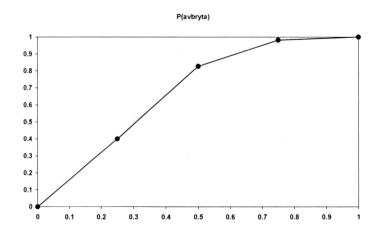

Index

adjusted R-square, 43, 73
AIC, 202
Akaikes information criterion, 154, 202
analysis of covariance, 88, 146
analysis of covariance model, 101
analysis of variance, 81, 146
analysis of variance table, 72
ANCOVA, 88
ANOVA, 81
Anova model, 101
ANOVA table, 15
assumptions in general linear models, 91
asymptotic relative efficiency, 110
attenuated, 248
attenuated correlation, 250
autoregressive, 198

backward elimination, 48
baseline measurements, 199
baseline model, 203
blinded, 242
Bonferroni, 85
Box-Cox transformations, 58
boxplot, 85, 186

canonical link, 147
canonical parameter, 147
case histories, 235
case-control studies, 236
class variables, 79

classification variables, 79
clinical trials , 233
coefficient of determination, 19, 42, 73
cohort studies, 236
completely randomized design, 241
compound symmetry, 191, 197
computer software, 91
conditional independence, 163
conditional odds ratio, 164
confidence interval, 74
constraints, 71, 99
contingency table, 157
contrast, 83
control group, 237
correlation, 19
count data, 157
counterfactual, 240
covariance analysis, 88
covariance structure, 197
cross-over design, 208
cross-over trials, 246
cross-sectional study, 236
cumulative logits, 170

dependent variable, 11, 70
design matrix, 98
determinant, 293
deterministic model, 70
double blind, 234, 237, 242
dummy variable, 79, 81

effect, 240

eigenvalue, 293
eigenvector, 294
eligibility, 237
endpoint, 237, 240
estimable functions, 90
estimated residual, 25, 41
expected frequencies, 158
exponential dispersion family, 147
exponential family, 147
exponential model, 56

F test, 73
factorial experiment, 85
Fisher's exact test, 119
fitted value, 25, 71
forecasting, 17
forward selection, 49
frequency table, 157
Friedmans test, 116

G-side effects, 195
Gamma distribution, 172
general linear model, 69, 70, 146
generalized inverse, 71, 99, 292
generalized linear mixed model, 208
generalized linear model, 146
GLM, 70
growth curve analysis, 187, 192

hat notation, 13, 71
homogeneity test, 132
homogeneous association, 163
homoscedasticity, 26, 91
humoral pathology, 234
hypergeometric distribution, 120

independent variable, 11, 70
individual profile plot, 185
inference on single parameters, 43
interaction, 85, 158
interaction plot, 87

intercept, 13, 70
intraclass correlation, 248
inverse, 291

Kruskal-Wallis test, 113

lack of fit sum of squares, 23
latent variable, 169
Latin square, 205
least squares, 13, 71
linear in the parameters, 53
linear predictor, 146
linear regression, 146
link function, 147
log-linear model, 146, 157, 158
logistic growth function, 60
logit analysis, 262
longitudinal data, 183

MANOVA, 187, 192
marginal odds ratio, 164
marginal probability, 157
matrix, 287
maximum likelihood, 71
mean profile plot, 185
Minitab, 28, 58, 75, 101
mixed model, 194
model, 12, 69
model building, 46, 92
model building strategy, 203
multinomial distribution, 160
multiple linear regression, 39, 77
multiple logistic regression, 153
multivariate analysis of variance, 187
mutual independence, 163

negative binomial distribution, 160
non-linear regression, 60
nonlinear models, 53
normal equations, 71, 99
normal probability plot, 25

observational studies, 235

observed residual, 71
odds, 152
odds ratio, 153, 162, 165, 253
offset, 167
oneway ANOVA, 12
ordinal logit regression, 171
ordinal probit regression, 171
ordinal regression, 171
over-all F test, 42
overdispersion, 160

pairwise comparison, 82
parallel groups designs, 241
parallel profiles, 193
partial odds ratio, 165
partial sum of squares, 75
Pearson chi-square, 162
Phase I , 239
Phase II, 239
Phase III, 239
Phase IV, 239
placebo, 242
placebo effect, 243
Poisson distribution, 159
Poisson regression models, 165
polynomial regression, 53
polynomials, 200
predicted value, 41, 71
prediction, 45
prediction interval, 18
probit analysis, 149, 262
probit models, 152
Proc Genmod, 150
Proc GLM, 83
Proc NLIN, 61
Proc REG, 53
proportional odds, 170, 171
protocol, 241
Pure Error sum of squares, 23

R-side effects, 195
R-square, 19, 42, 73
randomized clinical trial, 234

rank, 293
rank transformation, 117
rate data, 167
re-sampling, 122
regression analysis, 11
regression diagnostics, 25
regression model, 101
regression sum of squares, 16, 23, 42
reliability, 247
REML, 195
repeated measures, 183
repeated measures ANOVA, 186, 189
residual, 70, 71
residual sum of squares, 16, 42, 72
response variable, 11
restricted maximum likelihood, 195

SAS, 29, 75, 83, 101
saturated model, 159
sequential sum of squares, 75
sign test, 109, 115
simple linear regression, 11, 76
slope, 13
spatial power, 198
sphericity, 191
split-plot model, 190
square matrix, 288
statistical independence, 157
statistical model, 70
stepwise regression, 48
stepwise selection, 49
sum of squares, 72
summary measures, 186
surrogate endpoint, 240

t test, 79
testing linearity, 21, 30
tests on subsets of the parameters, 44, 74

ties, 112
total sum of squares, 16, 72
trace, 293
transform both sides, 63
transformations, 56
transpose, 288
type 1 SS, 48, 75
type 2 SS, 48, 75
type 3 SS, 75
type 4 SS, 76

unit vector, 289
unstructured, 197

validity, 247
variance components, 194
variance of b1, 17
vector, 287

Wilcoxon-Mann-Whitney's test, 111
within subject effect, 195